T0257651

Current Developments in Air Navigation Services

Current Developments in Air Navigation Services

Edited by **Margaret Ziegler**

CLANRYE INTERNATIONAL

New Jersey

Published by Clanrye International,
55 Van Reypen Street,
Jersey City, NJ 07306, USA
www.clanryeinternational.com

Current Developments in Air Navigation Services
Edited by Margaret Ziegler

International Standard Book Number: 978-1-63240-124-3 (Hardback)

Printed in the United States of America.

Contents

Preface

Every book is a source of knowledge and this one is no exception. The idea that led to the conceptualization of this book was the fact that the world is advancing rapidly; which makes it crucial to document the progress in every field. I am aware that a lot of data is already available, yet, there is a lot more to learn. Hence, I accepted the responsibility of editing this book and contributing my knowledge to the community.

Current developments in the field of air navigation services have been comprehensively described in this well-structured book. The division of air navigation services entered a new era of performance scheme that aims to meet requisite targets in four fundamental performance areas of safety, capacity, eco-friendliness and cost-efficiency. While it is highly expected that the targets are completely achieved, it is not recommended how; this being characteristic of performance based and goal oriented regulations. These key performance regions are interconnected by proportional and inversely proportional interrelations. For instance; if one aspires to expand sector capacity using the existing human resources (constant staff costs) but does not invest in technology (constant support cost) to accomplish higher cost-efficiency of service provision, the consequent overloaded system might unlock the Pandora's box of underlying safety issues. Since there is no scope of failure, we - the common, migrating and traveling public, airspace users, airport operators, air navigation services providers and the economy - will only benefit from this system; achieving goals of performance scheme in the process. However, the fundamental question which remains unanswered is "What is the winning strategy?" This book provides an understanding of noteworthy oddities of the elements of new business models of air navigation services provision as advancement of the latter became vital.

While editing this book, I had multiple visions for it. Then I finally narrowed down to make every chapter a sole standing text explaining a particular topic, so that they can be used independently. However, the umbrella subject sinews them into a common theme. This makes the book a unique platform of knowledge.

I would like to give the major credit of this book to the experts from every corner of the world, who took the time to share their expertise with us. Also, I owe the completion of this book to the never-ending support of my family, who supported me throughout the project.

Editor

Efficiency Assurance of Human-Centered and Technology Driven Air Traffic Management

Andrej Grebenšek

Additional information is available at the end of the chapter

1. Introduction

The European Air Navigation Services Providers (ANSPs) currently handle around 26,000 flights per day. This traffic should probably double by 2020. On average, traffic increased by roughly 7% per year until 2008. Following the global economic crisis, there has been a decline in traffic by 0.7% in 2009 and 0.4% in 2008 and afterwards again an increase of 8.7% in 2010 (European Commission, 2011).

However, there is also another side of the coin: the boom in air travel is exacerbating problems relating to the saturation levels reached at airports and the overloaded air traffic control (ATC) system. Airlines complain about the fragmentation of European airspace, which, they say, leads to inefficiency and major delays.

Europe enjoys a very high level of aviation safety. However, the constant increase in air traffic is putting pressure on safety, and this has consequences in terms of delays. The technical measures, taken to improve the management of airspace in recent years, have created additional capacity, but this has rapidly been outstripped by the growth of traffic. Passengers are now demanding a better quality of air transport service especially in terms of punctuality, given that it is no longer the exception that flights are over half an hour late.

The philosophy of Air Traffic Management (ATM) has not changed much since its beginning. Gradual increase in capacity of air traffic flows and airspace has only been achieved with the introduction of radar systems. On the other hand technology, methods and organization of work has still remained nearly the same. With present approach to solving the problems it is nearly impossible to achieve significant changes in quantity or quality of ATM.

Communication, navigation and surveillance domains improved and changed a lot over the last decade, thus enabling easier, faster and more precise navigation, direct routing of the aircraft and gradual transfer of separation responsibilities to the aircraft's cockpit. This will most probably lead to a leap to new technologies and organization of ATM.

New ATM strategy is now based on the "gate to gate" concept, including all phases of a flight. One major element of this strategy is that ATM system development has to be fully capacity driven in order to keep pace with the future demands of increasing air traffic. Another important item is the gradual transfer of responsibilities for separation between aircraft from ground ATC to aircraft themselves. Based on this strategy, a package of proposals has been designed by the European Commission, named Single European Sky (SES), granting political support to solving growing problems in the European sky (SESAR Joint Undertaking, 2009). Further on Single European Sky second package (SES II), made a significant step forward towards establishing targets in key areas of safety, network capacity, effectiveness and environmental impact (EUROCONTROL (EC-1), 2011).

In order to facilitate more efficient management of the European ATM, the Performance Review Commission (PRC), supported by the Performance Review Unit (PRU), has been established in 1998, under the umbrella of The European Organisation for the Safety of Air Navigation (EUROCONTROL). These entities introduced strong, transparent and independent performance review and target setting and provided a better basis for investment analyses and, with reference to existing practice, provided guidelines to States on economic regulation to assist them in carrying out their responsibilities (EUROCONTROL (EC-2), 2011).

Just recently, in December 2010, the European Commission adopted a decision which has set the EU-wide performance targets for the provision of air navigation services for the years 2012 to 2014. PRU ATM Cost-Effectiveness (ACE) benchmarking, has been recognized as one of the main inputs for determining the EU-wide cost-efficiency target and it will also have a major role in the assessment of national/FAB performance plans (EUROCONTROL, 2011).

Airspace users are putting constant pressure on the ANSPs, forcing them to improve their performance. Numerous airline associations call for urgent deliverables and a faster progress towards the Single European Sky (ATC Global INSIGHT, 2011). This all resulted in the initiative of the European Commission which is now setting the first priority on the Member States to revise their individual performance plans.

2. Background

Efficiency assurance can only be guaranteed through proper benchmarking of the current practices of Air Navigation Services provision. Commonly accepted tools for self-assessment have, among other, been the EUROCONTROL PRU ATM Cost-Effectiveness Benchmarking Report, which is, from 2002 on issued on a yearly basis, and to the smaller extend also Civil Air Navigation Services Organisation (CANSO) Global Air Navigation Services Performance Report, issued this year for the second time in the row (CANSO, 2011).

Both Reports are benchmarking similar issues, taking into account similar factors and similar variables. Major difference is though in the collection of ANSPs, where ACE Benchmarking Report focuses on all European actors and CANSO on global actors that volunteered to be benchmarked. Further on in this paper mainly ACE Benchmarking Report will be addressed.

For the purpose of this study it is assumed that Single European Sky packages I and II are defining the goals and targets and that ACE Benchmarking is broadly accepted tool for benchmarking.

The airspace covered by the SES and ACE Report is definite in size as well as traffic in the European airspace is constantly growing, but is again limited in the amount. Airspace users expect from ANSPs to have enough capacity to service their demand without any delay also in the peak periods of the day, month or year. The same expectation is shared by the general public and politicians. Delays are in general not accepted as they induce costs in excess of one billion euros per year.

For benchmarking purposes following KPIs have been set up by the PRU:

- Financial Cost-Effectiveness – The European ATM/CNS provision costs per composite flight hour with the sub-set of KPIs:
  - ATCO hour productivity – efficiency with which an ANSP utilizes the ATCO manpower;
  - ATCO employment costs per ATCO hour;
  - ATCO employment costs per Composite Flight Hour;
  - Support costs per Composite Flight Hour;
- Forward looking Cost-Effectiveness – forward looking plans and projections for the next five years;
- Economic Cost-Effectiveness, taking into account both financial cost-effectiveness and quality of service (ATFM ground delays, airborne holding, horizontal flight-efficiency and the resulting route length extension, vertical flight-efficiency and the resulting deviation from optimal vertical flight profile)

PRU recognizes both exogenous (factors outside the control of ANSP) and endogenous (factors entirely under the control of the ANSP) factors that can influence the ANSP performance.

This paper will in the remaining part focus on Financial Cost- Effectiveness, ATCO hour productivity and ATCO employment costs per ATCO hour and Composite Flight Hour

Significant volume of work has been done regarding the ATM Performance optimization. Some examples are listed under (Castelli et al., 2003; Castelli et al., 2005; Castelli et al., 2007; Christien et al., 2003; Fron, 1998; Kostiuk et al., 1997; Lenoir et al., 1997; Mihetec et al., 2011; Nero et al., 2007; Oussedik et al., 1998. Papavramides, 2009; Pomeret et al., 1997) and many more are available, however author of this paper was not able to find any paper that would challenge the benchmarking methodology.

According to the opinion of the author of this paper, different factors used in current benchmarking methodology, taking into the account also the assumptions above, can have a decisive effect on the objectivity of results of any benchmarking study and will therefore be further elaborated in the remaining part of this paper.

3. ACE benchmarking facts and figures

Overall financial cost-effectiveness (FCE) is one of the important parameters that are being benchmarked in the ATM Cost-Effectiveness (ACE) 2009 Benchmarking Report. Results are presented in Figure 1, presenting similar graph to the one in the above mentioned report.

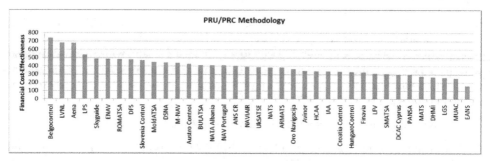

Figure 1. Overall financial cost-effectiveness 2009

Another output is graph on ATCO-hour Productivity (AHP), similar to the graph, presented in Figure 2.

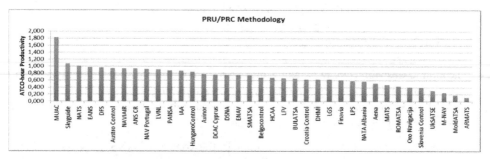

Figure 2. ATCO-Hour Productivity (gate-to-gate) 2009

Also important outputs are the ATCO employment costs per ATCO-hour (EC/AH) and ATCO employment costs per Composite Flight Hour (EC/CFH). Results are presented in Figure 3.

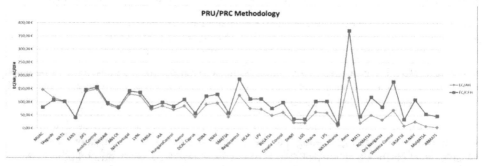

Figure 3. EC/AH and EC/CFH

If results of the calculations of the ratio of employment costs (EC) per CFH and employment costs per AH are compared across the full range of benchmarked ANSPs, trend becomes visible, showing that ANSPs with the employment costs per CHF and AH very close together are definitely much more efficient that the ones with the great difference between the two.

An ANSP to be efficient has to keep the EC per AH higher or equal to EC per CFH. EC can be eliminated from the equation, since on both sides of the formula they are the same. In order to achieve the above, CFH need to be higher or equal to the AH. This logic helps extracting the factors that are influencing the efficiency. The following formula proves that in order to enhance the efficiency an ANSP has to either increase the number of over flights or IFR airport movements or decrease the number of ATCOs or the number of their hours on duty:

$$EFH \ + \ (0.26 \, IAM) \geq N_{ATCOs} \ \bar{t}_{year} \tag{1}$$

This is easy to say but hard to do. En-route flight hours heavily depend on the geographical location, average overflying time, seasonal traffic variability etc. IFR airport movements mainly depend on the size of the airport which is closely linked to the attractiveness of the location and passenger's demand. Total number of ATCOs depends on required en-route and terminal capacity. That is again related to traffic demand, seasonal traffic variability, airspace complexity etc. Average ATCO-Hours on duty per ATCO per year are mainly a factor of social dialogue and legislation and are closely linked to the safety of operations.

4. Factors affecting the objectivity of benchmarking

4.1. ANSP size

ANSPs that are covered in PRU or CANSO report significantly vary per size and business. It is therefore hard to make an objective comparison of their performances. CANSO decided to group the ANSPs per number of IFR flight hours (see Table 1), but even within one group there are ANSPs that have at least twice the traffic than the other ones. Within the group A, the United States of America ANSP (FAA ATO) has twenty times more traffic than the Mexican ANSP (SENEAM). If we are to assume that the economy of scale contributes to the overall cost-effectiveness of the ANSPs then any type of comparison by pure facts only, cannot be considered as objective.

PRU clearly admits that their benchmarking is based purely on factual analysis and that many further factors would need to be considered in a normative analysis in order to make the results more comparable.

4.2. Traffic variability

ANSPs are by default expected to have enough capacity to match the demand of the airspace users at any period of the year, month, week or day. Especially for those performing scheduled flights delays induce costs that on the overall European level account for over a billion of euros per year. ANSPs therefore need to constantly enhance their capacity through upgrade of their technical facilities, technology and methods of work and by employment of additional staff, in particular ATCOs.

Grouping	ANSP	Total IFR Flight Hours
A (More than 1 million)	FAA ATO (USA)	25,106,283
	NAV CANADA	3,230,049
	AAI (India)	2,163,958
	NATS (UK)	1,731,274
	DFS (Germany)	1,366,637
	AENA (Spain)	1,358,390
	SENEAM (Mexico)	1,241,091
B (250,000 - 1 million)	NAV Portugal	468,728
	LFV (Sweden)	410,242
	Airways New Zealand	351,680
	AEROTHAI (Thailand)	320,360
	ROMATSA (Romania)	286,944
	ATNS (South Africa)	281,255
	IAA (Ireland)	256,550
C (100,000 - 250,000)	GCAA (UAE)	246,041
	ANS Czech Republic	231,079
	SMATSA (Serbia & Montenegro)	217,675
	NAVIAIR (Denmark)	209,917
	HungaroControl (Hungary)	197,909
	LVNL (The Netherlands)	151,592
	DCAC (Cyprus)	130,669
	Finavia (Finland)	114,645
D (0 - 100,000)	LPS (Slovak Republic)	82,382
	LGS (Latvia)	63,951
	NAATC (Netherlands Antilles)	55,623
	EANS (Estonia)	54,417
	Sloveniacontrol (Slovenia)	44,064
	Sakaeronavigatsia Ltd (Georgia)	42,590

Table 1. CANSO grouping of ANSPs per number of IFR Flight hours (CANSO. 2011)

This all adds "fixed" costs on a yearly basis, regardless of the actual demand in a particular period of the year, month, week or day. Due to the nature of business and required competency of the ANSPs staff, the personnel needed to cope with peak demand, usually in summer period, cannot be fired in October and re-employed in May next year. ANSPs rather need to keep them on their pay-roles throughout the whole year. The greater the variability of traffic the more the resources are underutilized and therefore contribute to cost ineffectiveness of a particular ANSP. So called "wasted resources" are presented in Figure 4 as a blue area.

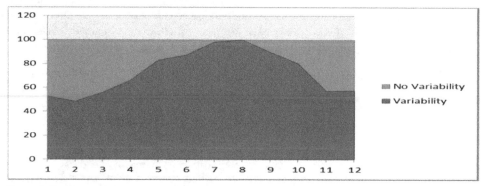

Figure 4. Traffic variability on a yearly basis

Airspace and traffic volumes are definite in size. It is simply not possible to optimize the business by purely attracting more traffic in the quiet periods of the year as firstly there is obviously no additional demand from the airspace users at those times and secondly, traffic flow can only be re-shifted at the expense of another ANSP. Traffic variability thus needs to be considered as a contributing factor that cannot be avoided.

PRU introduced seasonal traffic variability (TV) in their ATM Cost-Effectiveness (ACE) 2009 Benchmarking Report. It is calculated as ratio of traffic in the peak week (T_w) to the average weekly traffic ($\overline{T}$):

$$TV = \frac{T_w}{\overline{T}} \tag{2}$$

Calculated seasonal traffic variability factors for each ANSP are reported in the ATM Cost-Effectiveness (ACE) 2009 Benchmarking Report but are, to the knowledge of the author of this paper, only used to display the level of seasonal traffic variability at each particular ANSP and not directly used as corrective factors in the calculations.

The overall financial cost-effectiveness is calculated by a ratio of Air Traffic Management/Communication Navigation Surveillance (ATM/CNS) provision costs (ACPC) to the Composite flight hours:

$$FCE = \frac{ACPC}{CFH} \tag{3}$$

The ATM/CNS provision costs represent all costs of the ANSP for provision of the ATM/CNS service. Composite flight hours in (3) on the other hand are the summation of the En-route flight hours (EFH) and IFR airport movements (IAM) weighted by a factor that reflected the relative (monetary) importance of terminal and en-route costs in the cost base (EUROCONTOROL, 2011):

$$CFH = EFH + (0.26\,IAM) \tag{4}$$

The ATCO-hour Productivity is calculated by dividing Composite flight hours by Total ATCO-hours on duty:

$$AHP = \frac{CFH}{AH} \tag{5}$$

Where Total ATCO-hours on duty in (5) are the multiplication of Total number of ATCOs (N_{ATCOs}) and Average ATCO-Hours on duty per ATCO per year ($\bar{t}_{year}$):

$$AH = N_{ATCOs}\,\bar{t}_{year} \tag{6}$$

By using calculated seasonal traffic variability factors to equalize the composite flight hours using the formula below, the order of classification of the financial cost-effectiveness of the benchmarked ANSPs in Figure 1 changes. The ones with lower traffic variability move to the left towards less cost-effective ANSPs and the ones with higher traffic variability to the right, towards more cost-effective ANSPs.

$$Adjusted\ CFH = CFH \cdot TV \tag{7}$$

The same is valid also for the ATCO-Hour productivity presented in Figure 2.

4.3. Calculation of composite flight hours

CFH used for benchmarking by PRU/PRC are according to formula (4) composed of EFH and IAM weighted by a certain factor.

EFH are obtained from the EUROCONTROL statistical data and represent the amount of actual hours that flights, overflying particular area of responsibility of a particular ANSP, have spent in that particular portion of the airspace. The same figures can be obtained by multiplication of the number of flights (N_{of}) with the average overflying time of the relevant airspace ($\bar{t}_{of}$), using the formula below:

$$EFH = N_{of}\bar{t}_{of} \tag{8}$$

Average overflying time of European ANSPs ranges from roughly 10 minutes for the smallest ANSP to roughly 34 minutes for the ANSP which is lucky enough to have majority of the traffic along the longest routes in the route network. Looking at this time from another point of view means that if EFH is calculated in the PRU/PRC way, one single over flight attributes to 0,166 EFH for the smallest ANSP and on the other hand to 0,566 EFH for the ANSP with majority of the traffic along the longest routes. The difference is 3,4 times and means that the first ANSP would need to have at least 340% increase in traffic in order to match the productivity of the second ANSP, this all under the condition that the number of AH remains the same. There is no need to further elaborate that this is by no means possible.

On the other hand weight factor attributed to IAM translates to 0,26 CFH per single IFR airport movement, regardless whether the airport is a large national hub or a small regional airport.

Since terminal part of the CFH is calculated with the help of an artificial figure, equal for all ANSPs, regardless the size of the airport, it might be potentially wise to use the same logic also for the en-route part of the CFH, by simply attributing the weighted factor also to the EFH. This weighted factor could easily be the average calculated overflying time for all selected ANSPs.

5. Conclusion

ATM business does not always behave in line with the logic of the standard economy. It has its own set of legal rules, standards and recommended practices. On one hand everybody expects from it to be absolutely safe and efficient, but on the other hand airspace users constantly push for more financial efficiency expecting the business to be as cheap as possible. This could easily lead to contradiction.

By no doubt an ANSP has the power to influence certain factors that potentially influence the business, but there are other factors that have to be taken on board as a fact. This means that even if, by PRU standards more efficient ANSP takes over the so called less efficient ANSP, it would still have to overcome the same constraints or obstacles which are potentially effecting the business. This could also imply that if more efficient ANSP takes over the less efficient one, it does not immediately mean that now both ANSPs will become more efficient but would rather mean that the more efficient one would now become a bit less efficient.

Geographical position of an ANSP can be an advantage or an obstacle for the efficient performance. Seasonal traffic variability can attribute significantly to inefficiency of operations as airspace users pay for the full service 365 days in the year, but the ANSPs resources are only fully utilized in the peak period of the year. The calculated on average 25% of "wasted resources" per year, can potentially open a window of opportunity for optimization. Of course whole 25% could only be achieved in ideal conditions in a fictitious world, but on the other hand the European Commission asked the Member States to submit their Performance plans in such a manner that they will deliver incremental savings of only 3.5% per year for the SES II Performance Scheme reference period 2012 – 2014. Providing that only a portion of those 3.5% of savings is achieved through optimization of operations taking into account the seasonal traffic variability, it is already a step forward into the right direction.

The same goes for the calculation of the CFH. The proper solution to the problem could be in a design of a business model that would objectively support the managerial decision making processes. Until recently the business of the ANSPs has been regulated and full cost recovery regime allowed majority of the ANSP managers to only passively influence the business. On the other hand the new European Commission regulation introduces the requirements that would force everyone to optimize. Objective support in proper decision-making will therefore become essential.

When talking about ANSP performance it is mostly concluded that small ANSPs will most probably cease to exist and that the larger ones will take over their business. Looking at the graphs in Figures 2 and 3 this does not necessarily hold true as the Estonian ANSP even with the PRC/PRU methodology, easily compares with the German or U.K. ANSP. Obviously the parameters of the PRC/PRU benchmarking methodology somehow suit them. If proper corrections or adjustments are inserted in the benchmarking methodology more chance is given also to smaller or less trafficked ANSPs.

By using seasonal variability or different approach in calculations of the CFH the calculations addressing the performance of the ANSPs become a bit more objective. An ANSP that is situated in the geographical area with high seasonal traffic variability, could probably try to optimize as much as possible, but would hardly become more efficient than an ANSP with little seasonal traffic variability. On the other hand the CFH, the way they are calculated now definitely influence the results in some way. The methodology of calculations used by PRU/PRC favours, larger ones with a lot of terminal traffic.

This paper gives only one example on how methodology of calculations could potentially be improved. Proper benchmarking should foster proper decision-making. By improvements proposed the managerial decision-making process could be more adequately supported.

Author details

Andrej Grebenšek
University of Ljubljana, Faculty of Maritime Studies and Transport, Portorož, Slovenia

6. References

ATC Global INSIGHT. 2011. ATC Global Insight News. Available form internet: < http://www.atcglobalhub.com/ReadATMInsightNews.aspx?editid=newsid1015&titleid =editid96 >.

CANSO. 2011. *Global Air Navigation Services Performance Report 2011.*

EUROCONTROL . 2011. *ATM Cost-Effectiveness (ACE) 2009 Benchmarking Report.*

EUROCONTROL (EC-1). 2011. Single European Sky. Available from internet: < http://www.eurocontrol.int/dossiers/single-european-sky >.

EUROCONTROL (EC-2). 2011. Performance Review Commission. Available from internet: < http://www.eurocontrol.int/prc/public/subsite_homepage/homepage.html >.

European Commission. 2011. *Annual Analyses of the EU Air Transport Market 2010, Final Report.*

Castelli, L.; Omero, M.; Pesenti, R.; Ukovich, W. 2003. Evaluating the Performance of Air Control Centers. In *Proceedings of the 5ᵗʰ USA – EUROPE ATM R&D Seminar Budapest, Hungary.*

Castelli, L.; Ukovich, W.; Debels, P. 2005. Route Charging Policy for a Functional Block of Airspace (CEATS). In *Proceedings of the 6ᵗʰ USA – EUROPE ATM R&D Seminar, Baltimore, MD, USA.*

Castelli, L.; Ranieri, A. 2007. Air Navigation Service Charges in Europe. In *Proceedings of the 7th USA – EUROPE ATM R&D Seminar, Barcelona, Spain.*

Christien, R.; Benkouar, A. 2003. Air Traffic Complexity Indicators & ATC Sectors Classification. In *Proceedings of the 5ᵗʰ USA – EUROPE ATM R&D Seminar, Budapest, Hungary.*

Fron, X. 1998. ATM performance review in Europe. In *Proceedings of the 2ⁿᵈ USA – EUROPE ATM R&D Seminar, Orlando, FL, USA.*

Kostiuk, P. F.; Lee, D. A. 1997. Modeling the Capacity and Economic Effects of ATM Technology. In *Proceedings of the 1ˢᵗ USA – EUROPE ATM R&D Seminar, Saclay, France.*

Lenoir, N.; Hustache J-C. 1997. ATC Economic modeling. In *Proceedings of the 1ˢᵗ USA – EUROPE ATM R&D Seminar, Saclay, France.*

Mihetec, T.; Odić, D.; Steiner, S. 2011. Evaluation of Night Route Network on Flight Efficiency in Europe, International Journal for Traffic and Transport Engineering 1(3): 132 – 141.

Nero, G.; Portet, S. 2007. Five Years Experience in ATM Cost Benchmarking. In *Proceedings of the 7ᵗʰ USA – EUROPE ATM R&D Seminar, Barcelona, Spain.*

Oussedik, S.; Delahaye, D.; Schoenauer, M.1998. Air Traffic Management by Stohastic Optimization. In *Proceedings of the 2ⁿᵈ USA – EUROPE ATM R&D Seminar, Orlando, FL, USA.*

Papavramides, T. C. 2009. :"Nash equilibrium" situations among ATM Service Providers in Functional Airspace Bloks. A theoretical study. In *Proceedings of the Conference on Air Traffic Management (ATM) Economics, Belgrade, Serbia.*

Pomeret, J-M.; Malich, S. 1997. Piloting ATM Through Performance, In *Proceedings of the 1ˢᵗ USA – EUROPE ATM R&D Seminar, Saclay, France.*

SESAR Joint Undertaking. 2009. *European Air Traffic Management Master Plan, Edition 1.*

Chapter 2

Predicting Block Time:
An Application of Quantile Regression

Tony Diana

Additional information is available at the end of the chapter

1. Introduction

According to Merriam Webster[1], to predict is 'to declare or indicate in advance; foretell on the basis of observation, experience, or scientific reason.' The advent of sophisticated mathematical and statistical techniques has taken 'divination' out of prediction. In the late 19th century, the work of Francis Galton in the areas of regression analysis, correlation and the normal distribution has been instrumental in helping analysts investigate the relationship between dependent and independent variables and, as a result, to be able to improve forecast. More recently, economists such as Robert Engle and Clive Granger have made significant contributions to the study of time series that have widespread applications nowadays in economics and especially finance, such as price and interest rate volatility, as well as risk measurement.

Aviation is another industry that faces risk and uncertainty and has greatly benefited by advances in mathematical, statistical and operations research techniques. A flight is an event that can be scheduled up to six months ahead of its execution. However, despite the best preparation, flight performance is subject to many factors beyond human control such as weather, equipment failure, labor actions, security threats, etc. As a main contributor to the economy and global trade, government regulators, airlines and airport authorities have a vested interest in ensuring that the aviation system supports unimpeded movements of goods and people from their origin to their final destination. According to the *Total Delay Impact Study*[2] by a group of Nextor researchers, "the total cost of all US air transportation delays in 2007 was $32.9 billion. The $8.3 billion airline component consists of increased

---

[1] The source is http://www.merriam-webster.com/dictionary/predict.
[2] Ball, M. et al., 2010. *Total delay impact study, a comprehensive assessment of the costs and impacts of flight delay in the United States*, Nextor, vii. The report is available at the following website: http://its.berkeley.edu/sites/default/files/NEXTOR_TDI_Report_Final_October_2010.pdf.

expenses for crew, fuel, and maintenance, among others. The $16.7 billion passenger component is based on the passenger time lost due to schedule buffer, delayed flights, flight cancellations, and missed connections. The $3.9 billion cost from lost demand is an estimate of the welfare loss incurred by passengers who avoid air travel as the result of delays".

Predictability is all the more difficult to achieve as airlines often face three types of delay. First, delays can be induced: The air traffic control authority can initiate a ground delay program in case of adverse weather conditions or heavy traffic volume on the ground or en-route. Second, delays can be propagated: In a sequence of legs operated by the same tail-numbered aircraft, a flight may accumulate delays that cannot be recovered by the end of the itinerary. Finally, delays can be stochastic because they are the results of random events such as equipment breakdown or extreme weather events.

Predictability represents a key performance area in the aviation industry for several reasons.

- For the International Civil Aviation Organization (ICAO), predictability refers to the "ability of the airspace users and ATM service providers to provide consistent and dependable levels of performance."[3]
- One of the goals of the U.S. Next Generation of Air Transportation System (NextGen) is to foster the transition from an air traffic control to more of an air traffic managed system where pilots have more flexibility to select their routes, utilize performance-based navigation (PBN) with the help of satellites and make decisions based on automated information-sharing.
- According to Rapajic (2009:51), "cutting five minutes of average of 50 per cent of schedules thanks to higher predictability would be worth some €1,000 million per annum, through savings or better use of airlines and airport resources." Unpredictability imposes considerable costs on airlines in the forms of lost revenues, customer dissatisfaction and potential loss of market share.

Recently, much discussion has revolved around the validity of using airlines' schedules as a measure of on-time performance and the variance of block delay as an indicator of predictability. Both airlines' limited control over the three types of delay and airport congestion make it difficult to build robust schedules and to use schedule as a reference for on-time performance. In fact, schedule padding may skew actual airline performance assessment, hence the need for an alternative methodology.

This article proposes a methodology to determine the predictability of block time based on the case study of the Seattle-Oakland city pair. The predictable block time is located at the percentile where the sign and magnitude of the pseudo coefficient of determination is the highest, while all the covariates are significant at a given confidence level. Ordinary-least-square (OLS) regression models enable analysts to evaluate the percentage of variation in actual block time explained by changes in selected operational variables. However, quantile

---

[3] Henk J. Hof, Development of a Performance Framework in support of the Operational Concept, ICAO Mid Region Global ATM Operational Concept Training Seminar, Cairo, Egypt, November 28–December 1, 2005, p. 36.

regression is more robust to outliers than the traditional OLS regression because the latter does not focus on the conditional mean.

This is of importance to aviation practitioners and, especially, airline schedulers who have often resorted to schedule padding in order to make up for ground and en route delays. This research presents a different perspective on the study of predictability with the intent to help aviation analysts achieve the following objectives:

- To assess the impact of selected operational covariates at different locations of the distribution of block time.
- To derive more predictable block times based on the impact of operational covariates at various quantiles.
- To test a model without any assumption about the distribution of errors and homoscedasticity (constant variance of the residuals).

After a brief background, the discussion will proceed with the methodology, an explanation of the outcomes and some final comments.

2. Background

A focus group including communication navigation surveillance and air traffic management representatives[4] defined predictability as "a measure of delay variance against a performance dependability target. As the variance of expected delay increases, it becomes a very serious concern for airlines when developing and operating their schedules".

According to Donohue et al. (2001:398), "predictability focuses on the variation in the ATM [Air Traffic Management] system as experienced by the user. Predictability includes both variability in flight times and arrival rates". In this article, the study of predictability is extended beyond wheels-off (takeoff) and wheels-off (landing) times to include any flight operations between gate-out and gate-in times such as taxi-out and taxi-in movements. This approach takes into account passenger experience.

For Vossen et al. (2011:388), "flexibility can be defined as the amount of operational latitude granted to the carriers in meeting their individual objectives (e.g. on-time arrival, network preservation, profit) when disruptions occur. [...] The notion of predictability is closely related, and can be defined as the reduction of uncertainties in the implementation of ATFM [Air Traffic Flow Management] initiatives". Although airlines have to face many events in the course of a flight that cannot thoroughly be anticipated and planned for, "ATFM initiatives should provide the user with time to react, and the provider's intent should be communicated as clearly and as far in advance as possible".

Predictability is sometimes associated with the concept of robust airline scheduling. The latter is the outcome of four sequential tasks as schedule generation, fleet assignment,

---

[4] Report of the Air Traffic Services Performance Focus Group and Communication Navigation Surveillance, February 1999. *Airline Metric Concepts for Evaluating Air Traffic Service Performance*. The website is http://www.boeing.com/commercial/caft/cwg/ats_perf/ATSP_Feb1_Final.pdf.

aircraft routing and crew pairing/rostering (Wu 2010; Abdelghany and Abdelghany 2009). Fleet assignment models (FAM) are often used to determine how demand for air travel is met by available fleet (see Abara 1989 and Hane et al. 1995). Moreover, the fleet assignment models present two challenges: complexity and size of the problem that the FAM can handle.

Rapajic (2009) identified network structure and fleet composition as sources of flight irregularities. Wu (2010) provided an excellent exposition of issues related to delay management, operating process optimization, and schedule disruption management. Wu explained that "airline schedule planning is deeply rooted in economic principles and market forces, some of which are imposed and constrained by the operating environment of the [airline] industry" (2010:11). He presented a schedule optimization model to improve the robustness of airline scheduling. However, such a model does not consider how selective operational variables are likely to impact scheduling.

Morrisset and Odoni (2011) compared runway system capacity, air traffic delay, scheduling practices, and flight schedule reliability at thirty-four major airports in Europe and the United States from 2007 to 2008. The authors explained that European airports limit air traffic delay through slot control. The other difference is that declared capacity (therefore, the number of available slots) is based mainly on operations under instrument meteorological conditions (IMC). By not placing any restrictions on the number of operations, schedule reliability in the United States depends more on weather conditions than at European airports.

3. Methodology

3.1. The sample and the assumptions

The sample includes daily data for the month of June to August in 2000, 2004, 2010 and 2011 for the Seattle/Tacoma International (SEA)-Oakland International (OAK) city pair. The summer season is usually characterized by low ceiling and visibility that determine instrument meteorological conditions[5] and weather events such as thunderstorms—all likely to skew the distribution of block times.

Illustration 1 compares the boxplots of actual block times in minutes for the four summers under investigation. The boxplot shows the spread of the distribution, the selected quantile values, the position of the mean and median block times, and the presence of outliers that make it important to consider a regression model at different quantiles. The boxplots reveal an increase in the actual block times between summer 2004 and 2011. Summer 2010 features the largest range as well as the lowest block times at the 5th percentile among the four samples (Illustration 1). It is also characterized by the highest proportion of operations in instrument meteorological conditions compared with the other three samples (Table 1). The

---

[5] The minimum ceiling and visibility at SEA are respectively 4,000 feet and 3 nautical miles. At OAK, they are 2,500 feet and 8 nautical miles.

skewness coefficients[6] are 0.11, -0.44, 0.37, and 0.19 respectively for summer 2011, 2010, 2004 and 2000. A negative skew indicates that the left tail is longer. While the standard deviation is appropriate to measure the spread of a symmetric distribution, interquartile ranges are more indicative of spread changes in skewed distributions (see Figure 1).

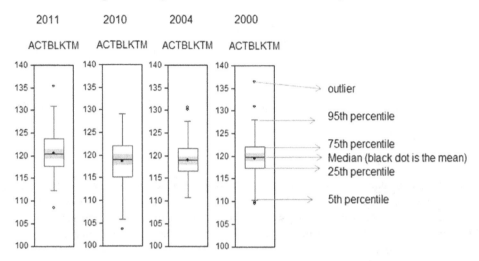

Figure 1. Box Plots of Actual Block Times (SEA-OAK)

Secondly, summer is part of the high travel season when demand is usually at its peak. This, in turn, is likely to increase airport congestion and subsequently impact block time. Finally, the years were selected to account for (1) pre- and post-September 11, 2001 traffic, (2) lower traffic demand resulting from the 2008-2009 economic recession, and (3) the introduction of the Green Skies over Seattle[7] after 2010.

The key performance indicators of flight performance are summarized in Table 1. Although the number of flights increased between 2000 and 2011 and the average minutes of expected departure clearance times (EDCT) were higher in 2011 than in 2000, the percentage of on-time gate departures and arrivals and other key delay indicators such as taxi-out delay (a measure of ground congestion) improved in 2011. It is interesting to point out that the percentage of flights in IMC did not change significantly at OAK among the four selected summers. IMC operations were, however, much higher in 2010 and 2011 than in 2000 at SEA, which explains the existence of average minutes of EDCT in 2010 and 2011.

The sample does not include a variable that measures performance-based navigation. The available surveillance data such as Traffic Flow Management System (TFMS) do not capture

---

[6] The skewness coefficient is computed as $\gamma = E[(x - \mu)^3/\sigma] = \mu^3/\sigma^3$ where μ^3 is the third moment about the mean μ and σ is the standard deviation and E is the expectation operator.

[7] The Green Skies over Seattle program includes initiatives such as reduced track mileage to minimum possible distance to protect the environment, optimized profile descent, reduction or elimination of low altitude radar vectoring, as well as required navigational performance.

whether a pilot had requested a required navigation performance procedure, whether air traffic control had granted the request, and whether the procedure had actually been implemented. Moreover, it is presently difficult to differentiate flown performance-based navigation procedures from instrument landing system (ILS) approaches in the case of flight track overlay.

SEA-OAK	Flight Count	% On-Time Gate Departures	% On-Time Airport Departures	% On-Time Gate Arrivals	Arrivals With EDCT*	Average EDCT Where EDCT>0	Gate Departure Delay (min)	Taxi Out Delay (min)	Average Taxi Out Time (min)	Airport Departure Delay (min)	Airborne Delay (min)	Taxi In Delay (min)	Block Delay (min)	Gate Arrival Delay (min)	Percent IMC** SEA	Percent IMC** OAK
2000	1,411	76.26	67.54	77.18	0	0	9.07	4.48	14.29	12.83	4.33	0.95	2.61	9.23	10.58	29.49
2004	1,473	73.52	68.30	79.16	0	0	11.03	3.49	12.80	13.84	6.93	1.22	1.44	9.26	8.33	29.71
2010	1,147	91.63	85.27	92.85	1	94	4.48	3.01	13.34	6.97	5.05	0.50	1.67	3.76	30.29	29.86
2011	1,196	93.73	90.64	94.90	1	104	3.83	2.79	13.03	5.92	3.92	0.70	1.90	3.24	22.83	29.78

* In the event of a ground delay, airlines are issued an expected departure clearance time (EDCT). Flights held by FAA at the departure airport due to problems at the arrival airport.
EDCT hold delay is computed by comparing EDCT wheels-off time to the flight plan's wheels-off time.
** Instrument Meteorological Conditions
Performance is compared with the last flight plan filed before wheels-off.
Source: ASPM

Table 1. Performance Metrics for the SEA-OAK City Pair

Secondly, the availability of Q-routes makes it possible for RNAV/RNP capable aircraft to reduce mileage, to minimize conflicts between routes and to maximize high-altitude airspace. Q-routes are available for use by RNAV/RNP capable aircraft between 18,000 feet MSL (mean sea level) and FL (flight level) 450 inclusive. They help minimize mileage and reduce conflicts between routes.

Thirdly, block time as a measure of gate-to-gate performance is sensitive to delays on the ground and en route. To account for this, airborne delay represents a surrogate for enroute congestion, while increases in taxi times imply surface movement congestion.

3.2. Sources and definition of the variables

The sources for the variables are ARINC[8]'s Out-Off-On-In times and the U.S. Federal Aviation Administration's Traffic Flow Management System (TFMS). The directional city pair data originated from the 'Enroute' and 'Individual Flights' data marts of the Aviation System Performance Metrics (ASPM) data warehouse[9].

The choice of variables reflects operational and statistical considerations. On the one hand, some model variables represent significant factors in airport congestion (taxi times) and enroute performance (airborne delays). On the other hand, the model with the highest values for the Akaike Information Criterion (AIC)[10] and Bayesian Information Criterion (BIC)[11] was selected in order to prevent overfitting and to reduce the number of covariates.

The dependent (response variable) and independent variables (covariates) are defined as follows:

---

[8] AIRINC stands for Aeronautical Radio, Inc. (http://www.arinc.com).
[9] The TFMS (formerly ETMS) and ARINC data as well as the ASPM delay metrics are available at http://aspm.faa.gov.
[10] The Akaike Information Criterion is defined as $2k - 2 \ln(L)$ where k is the number of parameters and L the maximized value of the likelihood function for the estimated model.
[11] The Bayesian Information Criterion is $-2 \ln(L) + k.\ln(n)$ where n is the number of observations.

- **Actual Block Time** (ACTBLKTM) is the dependent variable. It refers to the time in minutes from actual gate departure to actual gate arrival.
- **Block Buffer** (BLKBUFFER) represents the difference between planned and optimal block time. The latter is the sum of unimpeded taxi-out times and filed estimated time enroute. Block buffer is the additional minutes included in planned block time in order to take into account potential induced, propagated and stochastic delays. It has also been defined as "the additional time built into the schedule specifically to absorb delay whilst the aircraft is on the ground and to allow recovery between the rotations of aircraft" (Cook, 2007:105). Donohue et al. (2001:113) explained that "to obtain their desired on-time performance, airlines will add padding into a schedule to reflect an amount above average block times to allow for delay and seasonally experienced variations in block times."
- **Departure Delay** (DEPDEL) corresponds to difference between the actual and planned gate departure time at the departure airport in a city pair.
- **Arrival Delay** (ARRDEL) represents to the difference between the actual and planned gate arrival time at the arrival airport in a city pair.
- **Airborne Delay** (AIRBNDEL) accounts for the total minutes of airborne delay. It is the difference between the actual airborne times (landing minus takeoff times) minus the filed estimated time enroute.
- **Taxi-Out Time** (TXOUTTM) refers to the duration in minutes from gate departure to wheels-off times (gate-out to wheels-off).

3.3. Quantile regression

Readers interested in quantile regression are referred to Hao and Naiman (2007), Koenker (2005), Koenker and Hallock (2001) and Koenker and Bassett (1998). Quantile regression provides several advantages compared with the ordinary-least-square (OLS) regression in assessing the influence of selected operational factors on the variations of block time at various locations of its distribution:

- Quantile regression specifies the conditional quantile function and, therefore, a way to assess the probability of achieving a certain level of performance. It permits the analysis of the full conditional distributional properties of block time as opposed to ordinary-least-square (OLS) regression models that focus on the mean.
- It defines functional relations between variables for all portions of a probability distribution. Quantile regression can improve the predictive relationship between block times and selected variables by focusing on quantiles instead of the mean. As Hao and Naiman (2007:4) pointed out, "While the linear regression model specifies the changes in the conditional mean of the dependent variable associated with a change in the covariates, the quantile regression model specifies changes in the conditional quantile."
- It determines the effect of explanatory variables on the central or non-central location, scale, and shape of the distribution of block times.
- It is distribution-free, which allows the study of extreme quantiles. Outliers influence the length of the right tail and make average block time irrelevant as a standard for

identifying the best-possible block time. A single rate of change characterized by the slope of the OLS regression line cannot be representative of the relationship between an independent variable or covariate and the entire distribution of block time, the response variable. In the quantile regression, the estimates represent the rates of change conditional on adjusting for the effects of the other model variables at a specified percentile. Therefore, the skewed distribution of block times calls for a more robust regression method that takes into account outliers or the lack of sufficient data at a particular percentile (especially at the extremes of the distribution) and generates different slopes for different quantiles.

The difference between OLS and quantile regression characteristics are summarized in the table below:

Linear Regression	Quantile Regression
• Estimates the mean of a response variable conditional on the values of the explanatory variables (specifies the conditional mean function) • Determines the rate of change in the mean of the response variable	• Specifies the conditional quantile function (focus on quantiles). • Defines functional relations between variables for all portions of a probability distribution
• Provides a measure of the impact of explanatory variables on the central location of the distribution of the response variable. • Does not account for full conditional distribution properties of the response variable	• Determines the effect of explanatory variables on the central or non-central location, scale, and shape of the distribution of the response variable • Permits the analysis of the full conditional distributional properties of the dependent variable
Normal distribution (sensitive to outliers)	Distribution-free (allows study of extreme quantiles)
Determines best fitting line for all data	Different estimates for different quantiles
Normal distribution of errors	No assumption about the distribution of errors
Assumption of constant variance in errors (homoscedasticity)	Does not assume homoscedasticity

Table 2. The Assumptions of Linear and Quantile Regression

4. Outcomes and implications

Appendix 1 provides the estimates for the OLS models. The intercept that represents the predicted value of actual block time when the covariates are equal to zero is not significant at a 95% confidence level in the 2011 and 2010 samples. However, since the intercept is necessary to provide more accurate predictions, it was left in the model.

Among the dependent variables, gate arrival and departure delays are not significant at a 95% confidence level in the 2010 sample. This implies that airlines can make up for ground delays once en route or ground delays are likely to be more significant in a few extreme cases. The F statistics suggest that there is a zero percent chance that the dependent variable estimates are equal to zero. A value of the Durbin Watson statistic close to 2.00 suggests that there is little statistical evidence that the error terms are positively auto-correlated. The values of the coefficients of determination (R^2) imply that the model covariates explain a high proportion of the variation in block times.

In the quantile regression models, the covariate estimates, as well as the key regression statistics at the 5th, 25th, median, 75th and 95th percentile, are summarized in the appendix 2 table. The 50th quantile estimates can be used to track location changes. According to Hao and Naiman (2007: 55), the 5th and 95th percentiles "can be used to assess how a covariate predicts the conditional off-central locations as well as shape shifts of the response." In the case of the 50th percentile in summer 2011, the quantile regression model for at τ (tau) = 0.50 (50th percentile or median) is as follows:

$$\text{Block Time}_{\tau = 0.50} = -14.8091 - 0.8909 * X_{\text{BLKBUFFER}} + 1.0038 * X_{\text{SCHEDBLKTM}} - \tag{1}$$
$$0.3329 * X_{\text{DEPDEL}} + 0.2936 * X_{\text{ARRDEL}} + 1.0957 * X_{\text{AIRBNDEL}} + 1.1606 * X_{\text{TXOUTTM}} + \varepsilon$$

In equation (1), 1.1606 represents the change in the median of block time between SEA and OAK corresponding to a one minute change in taxi-out time at SEA. Since the p value is zero, we reject the null hypothesis, at a 95 percent confidence level, that taxi-out times at SEA has no effect on the median block time between SEA and OAK in summer 2011. The pseudo coefficient of determination is a goodness-of-fit measure[12]. In the case of summer 2011, 80.21% of the variation in block time is explained by the model covariates at the 50th percentile of block time (appendix 1).

No sample includes covariates that are significant at a 95% confidence level at all quantiles. Gate departure and arrival delays are significant only at the 95th percentile across the four samples. This means that departure and arrival delays are more likely to affect consistently block times in the upper percentiles—in case of severe airport congestion, for instance. Moreover, the magnitude of block buffers and gate departure delays have a negative impact on the conditional quantile of block time at all samples' selected percentiles. The size of the buffer and the time an aircraft will spend on the tarmac before take-off are conditions likely to affect block times. As a result, there is a need for analysts to decompose and to measure the different operations between gate-out and wheels-off times including gate departure, push-back, taxi-out and queuing times before wheels off. Airport Surface Detection

---

[12] See Koenker and Machado 1999 for further explanations. According to Fitzenberger et al. (2010: 234), "the pseudo R^2 equals one minus the sum of weighted deviations about estimated quantile over the sum of weighted deviations around raw quantile".

Equipment, Model X or ASDE-X data should help do so as the system relying on a combination of surface movement radar and transponder multi-lateration sensors becomes more widespread.

Taking the example of summer 2011, 95 percent of the distribution of block times between SEA and OAK was below 129.14 minutes compared with a mean of 120.62 minutes (appendix 2). In other words, there is a 95 percent chance that actual block time will be lower than 129.14 minutes—based on a quantile regression model that explains 85.34 percent of the variation in block times. One benefit of quantile regression is that it facilitates the evaluation of scale and magnitude changes across samples and percentiles.

The quantile regression estimates in appendix 1 imply that block times increased in between summer 2000 and 2011 at all quantiles. In a comparison of summer 2000 with summer 2011, there had been an increase of 2.21 minutes in block times at the 95th percentile, for instance. The SEA-OAK city pair has been mainly operated by Southwest Airlines (SWA) and Alaska Airlines (ASA) with a predominant fleet of Boeing 737s. The total number of ASA arrivals and departures declined to 356 in summer 2011 from 693 in summer 2000—with 91 ASA flights operated by Horizon's Bombardier Q400[13]. Nevertheless, ASA operated larger capacity models such as the dash 400, 800 and 900 series, while SWA utilized a combination of dash 300, 500 and 700 models. The reason for the increase in block time may be attributed to airlines' operations policy to slow aircraft speed in order to save on fuel costs[14]. Weather conditions characterized by the percentage of operations in instrument meteorological conditions (IMC) did not vary substantially at OAK compared with SEA (see Table 1).

In appendix 3, the graphs illustrate the 95% confidence bands in the case of summer 2000. The estimates show a positive relationship between the quantile value and the estimated coefficients for scheduled block times, taxi out times and airborne delay, with a stronger effect in the upper tail. The effect of gate departure and arrival delays is not relatively constant, especially at the 50th percentile as implied by the wider bands around the 50th percentile value. These graphs are important for the analysts in identifying the quantiles where quantile value is likely to be close to the estimated coefficients and, therefore, to improve the accuracy of predicted block time.

5. Final comments

Based on the analysis of the SEA-OAK city pair case study, this research showed how quantile regression can help aviation practitioners develop more robust schedules. Originally proposed by Koenker and Bassett (1978), quantile regression is a rather novel approach to the analysis of airlines' on-time performance. Although it is more widely used

---

[13] The sources for schedules and aircraft mix are the Official Airline Guide (http://www.oag.com) and Innovata (http://www.innovata-llc.com).

[14] Associated Press. *Airlines slow down flights to save on fuel: JetBlue adds 2 minutes to each flight, saves $13.6 million a year in jet fuel*, May 1, 2008. The article is available at the following website: http://www.msnbc.msn.com/id/24410809/ns/business-us_business/t/airlines-slow-down-flights-save-fuel/#.T01rmPES2Ag

in ecology and biology than in the transportation industry, quantile regression is seldom featured in econometric textbooks. Nevertheless, it presents several advantages.

First, it enables aviation analysts to consider the impact of selected covariates on different locations of the distribution of block times. Secondly, the significance and the strength of the impact of selected covariates on block times make it possible to assess the probability that gate-to-gate operations is likely to reach a specific duration. This is made possible by looking at the conditional quantile in the case of quantile regression as opposed to the conditional mean of the distribution of block times in the case of OLS models. Thirdly, quantile regression makes it easier to evaluate the scale and magnitude of change across specific percentiles over a sample. Finally, quantile regression can help analysts study the impact of covariates from different perspectives. For instance, in summer 2011, the data analysis suggests that 95% of the block time distribution will be below the quantile dependent variable value of 129.14 minutes as a result of the impact of the covariates' impact. Quantile regression enables the identification of more realistic threshold times based on quantiles and it allows airline practitioners to simulate and to evaluate various scenarios linked to changes in the models' covariates.

Predictability is a key performance area identified by the International Civil Aviation Organization. Moreover, it is a corner stone of the Next Generation of Air Transport System (NextGen) initiatives in the U.S. and the Single European Sky ATM Research program (SESAR) to ensure the transition from an air traffic controlled to a more air traffic managed environment. As air transportation regulators are under public pressure to crack down on tarmac and other types of delays, it has become imperative for airline schedulers to evaluate models that reflect the predictable influence of key operational variables on actual on-time performance. The complexity of the air traffic system, the inability for airline schedulers to fully anticipate both airport and en route congestion, and delays all make it more significant for aviation practitioners to assess the impact of some key operational variables at different locations of the distribution of block times that usually tends to be skewed due to outliers.

The imbalance between air travel demand and airport capacity usually results in delays. As block times become more predictable, it is more possible for airline and airport operators to optimize airport capacity— especially at large congested airports. This is all the more significant in the U.S. where arrival and departure flows are not slot- constrained as in Europe. Block time predictability does not only affect how airports and airlines operate, but also the capability of air traffic control authorities to anticipate staff workload, as well as the ability of ground handlers to minimize aircraft turn times by allocating resources where and when needed.

Author details

Tony Diana[15]
Division Manager, NextGen Performance, Federal Aviation Administration, Office of NextGen Performance and Outreach, ANG-F1, SW, Washington DC, USA

---

[15] Note: This article does not represent the opinion of the Federal Aviation Administration.

Appendix

The ordinary-least square regression outputs

Alpha = .95	2011		2010		2004		2000	
	Coefficient	Probability	Coefficient	Probability	Coefficient	Probability	Coefficient	Probability
ACTBLKTM (dep var)								
INTERCEPT	-3.3081	0.7037	-6.0440	0.3358	-14.3526	0.0154	-19.6283	0.0249
BLKBUFFER	-0.8383	0.0000	-0.9862	0.0000	-0.9051	0.0000	-0.8087	0.0000
SCHEDBLKTM	0.9291	0.0000	0.9552	0.0000	1.0531	0.0000	1.0940	0.0000
DEPDEL	-0.4482	0.0000	-0.0878	0.1046	-0.1263	0.0159	-0.2544	0.0000
ARRDEL	0.4319	0.0002	0.0726	0.2134	0.1399	0.0144	0.2554	0.0000
AIRBNDEL	0.9596	0.0000	1.0619	0.0000	0.8741	0.0000	0.9249	0.0000
TXOUTTM	0.9831	0.0000	1.0326	0.0000	0.8850	0.0000	0.7643	0.0000
R-squared	0.9472		0.9879		0.9819		0.9853	
Adjusted R-squared	0.9435		0.9871		0.9806		0.9843	
S.E. of regression	1.1578		0.5894		0.5681		0.6043	
Sum squared resid	113.9466		29.5253		27.4320		31.0360	
Log likelihood	-140.3837		-78.2615		-74.8788		-80.5568	
F-statistic	254.2718		1158.3370		767.2133		951.7874	
Prob(F-statistic)	0.0000		0.0000		0.0000		0.0000	
Mean dep var	120.6251		118.6857		119.1140		119.4768	
Durbin Watson	1.9670		1.8550		1.9048		2.0154	

Not significant at α = .95

Summer 2000: Quantile process estimates (95% confidence level)

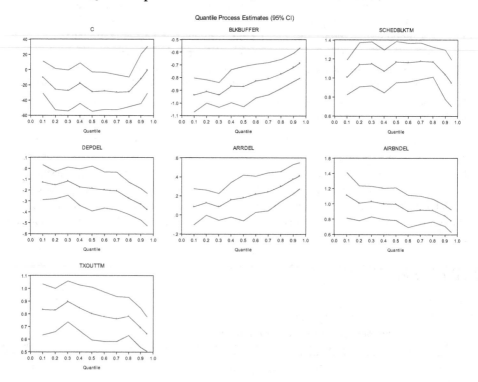

The quantile regression outputs

Alpha = .95	2011		2010		2004		2000	
	Coefficient	Probability	Coefficient	Probability	Coefficient	Probability	Coefficient	Probability
5th Percentile								
INTERCEPT	12.5331	0.2650	4.2506	0.7100	-4.6586	0.3253	-10.3640	0.2707
BLKBUFFER	-0.7264	0.0000	-1.0038	0.0000	-0.9720	0.0000	-0.9798	0.0000
SCHEDBLKTM	0.8110	0.0000	0.8646	0.0000	0.9664	0.0000	1.0065	0.0000
DEPDEL	-0.7312	0.0000	-0.0540	0.2859	-0.0339	0.6064	-0.0931	0.2397
ARRDEL	0.8353	0.0000	0.1076	0.0530	0.0493	0.5027	0.0357	0.7014
AIRBNDEL	0.8453	0.0000	1.0346	0.0000	0.9057	0.0000	1.1799	0.0000
TXOUTTM	0.6581	0.0074	1.0355	0.0000	0.9542	0.0000	0.8833	0.0000
Pseudo R-squared	0.7082		0.8896		0.8900		0.8791	
Adjusted R-squared	0.6876		0.8818		0.8822		0.8706	
S.E. of regression	2.7301		1.1423		0.9829		1.1722	
Quantile dependent var	112.8600		107.5600		113.0000		110.9400	
Mean dependent var	120.6251		118.6857		119.114		119.4768	
25th Percentile								
INTERCEPT	0.7081	0.9534	-10.3697	0.2900	-7.3948	0.1315	-22.7559	0.1164
BLKBUFFER	-0.8557	0.0000	-1.0066	0.0000	-0.9266	0.0000	-0.9058	0.0000
SCHEDBLKTM	0.8868	0.0000	0.9861	0.0000	0.9915	0.0000	1.1157	0.0000
DEPDEL	-0.4349	0.0003	-0.0351	0.5664	-0.1217	0.1107	-0.1446	0.0394
ARRDEL	0.3957	0.0004	0.0367	0.6397	0.1257	0.1233	0.1260	0.0862
AIRBNDEL	1.1110	0.0000	1.0671	0.0000	0.9263	0.0000	1.0099	0.0000
TXOUTTM	1.0340	0.0000	1.0522	0.0000	0.9112	0.0000	0.8284	0.0000
Pseudo R-squared	0.7694		0.8972		0.8799		0.8878	
Adjusted R-squared	0.7532		0.8899		0.8714		0.8799	
S.E. of regression	1.2129		0.7105		0.6992		0.7683	
Quantile dependent var	117.7100		115.0000		116.5300		117.2100	
Mean dependent var	120.6251		118.6857		119.114		119.4768	
50th Percentile								
INTERCEPT	-14.8091	0.5037	-12.1850	0.1575	-12.3493	0.0406	-28.8452	0.0316
BLKBUFFER	-0.8909	0.0000	-0.9754	0.0000	-0.9289	0.0000	-0.8694	0.0000
SCHEDBLKTM	1.0038	0.0000	0.9990	0.0000	1.0427	0.0000	1.1690	0.0000
DEPDEL	-0.3329	0.0366	-0.0560	0.3936	-0.1077	0.1146	-0.1849	0.0828
ARRDEL	0.2936	0.0333	0.0207	0.7928	0.1070	0.1486	0.1794	0.1486
AIRBNDEL	1.0957	0.0000	1.0409	0.0000	0.8574	0.0000	0.9972	0.0000
TXOUTTM	1.1606	0.0000	1.0893	0.0000	0.8686	0.0000	0.8021	0.0000
Pseudo R-squared	0.8021		0.8946		0.8674		0.8736	
Adjusted R-squared	0.7881		0.8871		0.8580		0.8647	
S.E. of regression	1.1988		0.6096		0.5835		0.6142	
Quantile dependent var	120.4300		119.0000		118.9300		119.8600	
Mean dependent var	120.6251		118.6857		119.114		119.4768	
75th Percentile								
INTERCEPT	-31.1458	0.0001	-2.2100	0.7869	-20.9044	0.0037	-25.7729	0.0169
BLKBUFFER	-0.8502	0.0000	-0.9943	0.0000	-0.8452	0.0000	-0.7926	0.0000
SCHEDBLKTM	1.1481	0.0000	0.9310	0.0000	1.1098	0.0000	1.1441	0.0000
DEPDEL	-0.3207	0.0512	-0.0846	0.3845	-0.1746	0.1296	-0.2313	0.0122
ARRDEL	0.3281	0.0642	0.0536	0.6036	0.1805	0.1081	0.2584	0.0127
AIRBNDEL	0.9996	0.0000	0.9994	0.0000	0.8917	0.0000	0.9390	0.0000
TXOUTTM	1.0485	0.0000	1.0311	0.0000	0.8113	0.0000	0.7669	0.0000
Pseudo R-squared	0.8269		0.8958		0.8639		0.8781	
Adjusted R-squared	0.8147		0.8884		0.8543		0.8695	
S.E. of regression	1.4266		0.7151		0.7214		0.7646	
Quantile dependent var	123.8000		121.9200		121.6000		122.1500	
Mean dependent var	120.6251		118.6857		119.114		119.4768	
95th Percentile								
INTERCEPT	-17.6714	0.2819	-17.4339	0.0149	-31.0322	0.0000	-0.2942	0.9850
BLKBUFFER	-0.9532	0.0000	-0.9099	0.0000	-0.7827	0.0000	-0.6847	0.0000
SCHEDBLKTM	1.0483	0.0000	1.0580	0.0000	1.1988	0.0000	0.9482	0.0000
DEPDEL	-0.2748	0.0332	-0.2929	0.0002	-0.3229	0.0003	-0.3772	0.0000
ARRDEL	0.3277	0.0174	0.2663	0.0018	0.3269	0.0006	0.4122	0.0000
AIRBNDEL	0.7805	0.0000	1.0384	0.0000	0.7939	0.0000	0.7791	0.0000
TXOUTTM	1.1530	0.0000	0.9544	0.0000	0.7826	0.0000	0.6418	0.0000
Pseudo R-squared	0.8534		0.8997		0.8859		0.9103	
Adjusted R-squared	0.8431		0.8926		0.8778		0.9040	
S.E. of regression	1.8510		1.0918		1.1257		1.1419	
Quantile dependent var	129.1400		127.4300		126.0600		126.9300	
Mean dependent var	120.6251		118.6857		119.114		119.4768	

Not significant at α = .95

6. References

Abara, J., 1989. Applying integer linear programming to the fleet assignment problem. Interfaces 19(4), 20-28.

Abdelghany, F. and Abdelghany, K., 2009. Modeling applications in the airline industry. Ashgate Publishing Company: Burlington, Vermont.

Cook, A., 2007. European air traffic management: principles, practices, and research. Ashgate Publishing Company: Burlington, Vermont.

Donohue, G.L., Zellweger, A., Rediess, H., and Pusch, C., 2001. Air transportation system engineering: progress in astronautics and aeronautics. American Institute of Aeronautics and Astronautics: Danvers, Massachusetts.

Fitzenberger, B., Koenker, R., and Machado, J., 2010. Economic applications of quantile regression (Studies in empirical economics). Physica Verlag: Heidelberg.

Hane, C.A., Barnhart, C., Johnson, E.L., Marsten, R.E., Nemhauser, G.L., and Sigismondi, G., 1995. The fleet assignment problem: solving a large scale integer programming. Mathematical Programming, 70(2), 211-232.

Hao, L. and Naiman, D. Q., 2007. Quantile Regression. Sage Publications: Thousand Oaks, CA.

Koenker, R., 2005. Quantile Regression. Cambridge, UK: Cambridge University Press.

Koenker, R. and Hallock K., 2001. Quantile Regression. Journal of Economic Perspectives, 15(4), 143-156.

Koenker, R. and Machado, J., 1999. Goodness of fit and related inference processes for quantile regression, Journal of the American Statistical Association, 94(448), 1296-1310.

Morissett, T. and Odoni, A., 2011. Capacity, delay, and schedule reliability at major airports in Europe and the United States. Transportation Research Record: Journal of the Transportation Research Board, 2214, 85-93.

Rapajic, J., 2009. Beyond airline disruptions. Ashgate Publishing Company: Burlington, Vermont.

Vossen, T., Hoffman, R., and Mukherjee, A., 2011. Air traffic flow management in Quantitative problem solving methods in the airline industry: a modeling methodology handbook, Barnhart, C. and Smith, B., editors. Springer: New York.

Wu, C.-L., 2010. Airline operations and delay management: insights from airline economics, networks and strategic schedule planning. Ashgate Publishing Company: Burlington, Vermont.

Chapter 3

Measuring Sector Complexity: Solution Space-Based Method

S.M.B. Abdul Rahman, C. Borst, M. Mulder and M.M. van Paassen

Additional information is available at the end of the chapter

1. Introduction

In Air Traffic Control (ATC), controller workload has been an important topic of research. Many studies have been conducted in the past to uncover the art of evaluating workload. Many of which have been centered on the sector complexity or task demand based studies [1,2,3,4]. Moreover, all have the aim to understand the workload that was imposed on the controller and the extent to which the workload can be measured.

With the growth in world passenger traffic of 4.8% annually, the volume of air traffic is expected to double in no more than 15 years [5]. Although more and more aspects of air transportation are being automated, the task of supervising air traffic is still performed by human controllers with limited assistance from automated tools and is therefore limited by human performance constraints [6]. The rise in air traffic leads to a rise in the Air Traffic Controller (ATCO) task load and in the end the ATCO's workload itself.

The 2010 Annual Safety Review report by European Aviation Safety Agency (EASA) [7] indicates that since 2006, the number of air traffic incidents with direct or indirect Air Traffic Management (ATM) contribution has decreased. However, the total number of major and serious incidents is increasing, with incidents related to separation minima infringements bearing the largest proportion. This category refers to occurrences in which the defined minimum separation between aircraft has been lost. With the growth of air traffic, combined with the increase of incidents relating to separation minima infringements, a serious thought have to be put into investigating the causes of the incidents and plans on how to solve them.

Initiatives to design future ATM concepts have been addressed in both Europe and the United States, within the framework of Single European Sky ATM Research (SESAR) [8] and Next Generation Air Transportation System (NextGen) [9]. An increased reliance on airborne and ground-based automated support tools is anticipated in the future ATM concept by SESAR and NEXTGEN. It is also anticipated that in both SESAR and NEXTGEN concepts a better management of human workload will be achieved. However, to enable that, a more comprehensive understanding of human workload is required, especially that of controllers.

This chapter wil start with a discussion on sector complexity and workload and is followed by a deliberation of previous and current sector complexity and workload measures. Next, a method called the Solution Space Diagram (SSD) is proposed as a sector complexity measure. Using the SSD, the possibility of measuring different sector design parameters are elaborated and future implications will be discussed.

2. Sector complexity and workload

ATCO workload is cited as one of the factors that limit the growth of air traffic worldwide [10,11,12]. Thus, in order to maintain a safe and expeditious flow of traffic, it is important that the taskload and workload that is imposed on the ATCO is optimal. In the effort to distinguish between taskload and workload, Hilburn and Jorna [1] have defined that system factors such as airspace demands, interface demands and other task demands contribute to task load, while operator factors like skill, strategy, experience and so on determine workload. This can be observed from Figure 1.

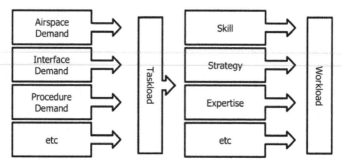

Figure 1. Taskload and Workload Relation [1].

ATCOs are subject to multiple task demand loads or taskloads over time. Their performance is influenced by the intensity of the task or demands they have to handle. Higher demands in a task will relate to a better performance. However, a demand that is too high or too low will lead to performance degradation. Thus, it is important that the demand is acceptable to achieve optimal performance.

The workload or mental workload can be assessed using a few methods such as using performance-based workload assessment through primary and secondary task performance, or using subjective workload assessment through continuous and discrete workload ratings, and lastly using physiological measures. However, because physiological measures are less convenient to use than performance and subjective measures, and it is generally difficult to distinguish between workload, stress and general arousal, these are not widely used in assessing workload [13].

Previous studies have also indicated that incidents where separation violations occurred can happen even when the ATCO's workload is described as moderate [14,15]. These incidents can be induced by other factors such as inappropriate sector design. Sector design is one of the key components in the airspace complexity. Airspace complexity depends on both structural and flow characteristics of the airspace. The structural characteristics are fixed for a sector, and depend on the spatial and physical attributes of the sector such as terrain, number of airways, airway crossings and navigation aids. The flow characteristics vary as a function of time and depend on features like number of aircraft, mix of aircraft, weather, separation between aircraft, closing rates, aircraft speeds and flow restrictions. A combination of these structural and flow parameters influences the controller workload [16].

A good airspace design improves safety by avoiding high workload for the controller and at the same time promotes an efficient flow of traffic within the airspace. In order to have a good airspace design, the ATC impact of complexity variables on controller workload has to be assessed. Much effort has been made to understand airspace complexity in order to measure or predict the controller's workload. In this chapter the solution space approach is adopted, to analyze in a systematic fashion how sector designs may have an impact on airspace complexity, and ultimately the controller workload.

2.1. Previous research on complexity factors

The Air Traffic Management (ATM) system provides services for safe and efficient aircraft operations. A fundamental function of ATM is monitoring and mitigating mismatches between air traffic demand and airspace capacity. In order to better assess airspace complexity, methods such as 'complexity maps' and the 'solution space' have been proposed in Lee et. al [17] and Hermes et al. [18]. Both solutions act as an airspace complexity measure method, where a complexity map details the control activity as a function of the parameters describing the disturbances, and the solution space details the two-dimensional speed and heading possibilities of one controlled aircraft that will not induce separation violations.

Much effort has been made to understand airspace complexity in order to measure the controllers' workload. Before introducing the solution space approach, first some more common techniques are briefly discussed.

2.1.1. Static density

One of the methods to measure complexity is the measurement of aircraft density and it is one of the measures that are commonly used to have instant indication of the sector complexity. It is defined as the number of aircraft per unit of sector volume. Experiments indicated that, of all the individual sector characteristics, aircraft density showed the largest correlation with ATCO subjective workload ratings [19,20]. However, aircraft density has significant shortcomings in its ability to accurately measure and predict sector level complexity [19,21]. This method is unable to illustrate sufficiently the dynamics of the behavior of aircraft in the sector. Figure 2 shows an example where eight aircraft flying in the same direction do not exhibit the same complexity rating when compared to the same number of aircraft flying with various directions [18].

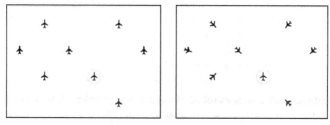

Figure 2. Example of Different Air Traffic Orientation.

2.1.2. Dynamic density

Another measurement of sector complexity is dynamic density. This is defined as "the collective effort of all factors or variables that contribute to sector-level ATC complexity or difficulty at any point of time" [19]. Research on dynamic density by Laudeman et al. [22] and Sridhar et al. [16] has indicated few variables for dynamic density and each factor is given a subjective weight. Characteristics that are considered include, but not limited to the number of aircraft, the number of aircraft with heading change greater than 15° or speed change greater than 10 knots, the sector size, and etc. The calculation to measure dynamic density can be seen in Equation (1).

$$\text{Dynamic Density} = \sum_{i=1}^{n} W_i DV_i \tag{1}$$

where dynamic density is a summation of the Dynamic Variable (DV) and its corresponding subjective weight (W). The calculation of the dynamic density is basically based on the weights gathered from regression methods on samples of traffic data and comparing them to subjective workload ratings. Essentially, the assignment of weights based on regression methods means that the complexity analysis based on dynamic density could only be performed on scenarios that differ slightly from the baseline scenario. Therefore the metric is not generally applicable to just any situation [18].

2.1.3. Solution space-based approach

Previous work has shown that the SSD is a promising indicator of sector complexity, in which the Solution Space-based metric was proven to be a more objective and scenario-independent metric than the number of aircraft [18,23,24]. The Forbidden Beam Zone (FBZ) of Van Dam et al. [25] has been the basis for representing the SSD. It is based on analyzing conflicts between aircraft in the relative velocity plane. Figure 3 (a) shows two aircraft, the controlled aircraft (A_{con}) and the observed aircraft (A_{obs}). In this diagram, the protected zone (PZ) of the observed aircraft is shown as a circle with radius of 5NM (the common separation distance) centered on the observed aircraft. Intrusion of this zone is called a 'conflict', or, 'loss of separation'. Two tangent lines to the left and right sides of the PZ of the observed aircraft are drawn towards the controlled aircraft. The area inside these tangent lines is called the FBZ.

This potential conflict can be presented on a SSD. Figure 3 (b) shows the FBZ in the SSD of the controlled aircraft. The inner and outer circles represent the velocity limits of the controlled aircraft. Now, if the controlled aircraft velocity lies inside the triangular-shaped area, it means that the aircraft is headed toward the PZ of the observed aircraft, will eventually enter it, and separation will be lost.

The exploration of sector complexity effects on the Solution Space parameters and, moreover, workload is important in order to truly understand how workload was imposed on controllers based on the criteria of the sector. Having the hypotheses that sector parameters will have a direct effect on the SSD geometrical properties, the possibility of using the SSD in sector planning seems promising. Figure 4 shows the relationship between taskload and workload as described by Hilburn and Jorna [1], where we adapted the position of sector complexity within the diagram. The function of the SSD is included as a workload measure [18,23,24] and alleviator [26] and also the possibility of aiding sector planning through SSD being a sector complexity measure [24].

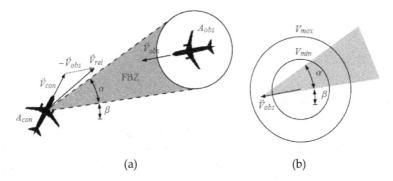

(a)   (b)

Figure 3. Two Aircraft Condition (a) Plan View of Conflict and the FBZ Definition. (b) Basic SSD for the Controlled Aircraft. (Adapted from Mercado-Velasco et al., [26])

Initial work by Van Dam et al. [25] has introduced the application of the Solution Space in aircraft separation problems from a pilot's perspective. Hermes et al. [18], d'Engelbronner et al. [23], Mercado-Velasco et al. [26] and Abdul Rahman et al. [24] have transferred the idea of using the Solution Space in aircraft separation problems for ATC. Based on previous research conducted, a high correlation was found to exist between the Solution Space and ATCO's workload [18,23,24]. Abdul Rahman et al. [24] also investigated the possibility of measuring the effect of aircraft proximity and the number of streams on controller workload using the SSD and have discovered identical trends in subjective workload and the SSD area properties. Mercado-Velasco et al. [26] study the workload from a different perspective, looking at the possibility of using the SSD as an interface to reduce the controller's workload. Based on his studies, he indicated that the diagram could indeed reduce the controller's workload in a situation of increased traffic level [26].

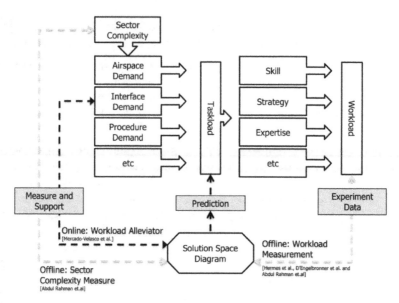

Figure 4. Solution Space Diagram in Measuring and Alleviating Workload (adapted from Hilburn and Jorna [1])

3. Complexity measure using the solution space diagram

The results gathered here are based on offline simulations of more than 100 case studies with various situations as detailed in this chapter. The affected SSD area has been investigated to understand the effects of sector complexities on the available solution space. Conclusions from previous work by Hermes et al. [18] and d'Engelbronner et al. [23] stated that the available area in the Solution Space that offers solutions has a strong (inverse) correlation with ATCO workload. In this case study, two area properties were investigated in order to measure the complexity construct of the situation, which are the total area affected (A_{total}) and the mean area affected (A_{mean}) for the whole sector. The A_{total} percentage is the area covered by the FBZs as a percentage of the total area between the minimum and the maximum velocity circles in the SSD, based on the currently controlled aircraft. The A_{mean} percentage affected is the A_{total} affected for all aircraft in the sector divided by the number of aircraft. This will give an overview of the complexity metric for the whole sector.

$$A_{total} = \sum A_{affected} \tag{2}$$

$$A_{mean} = \frac{\sum_{t=1}^{n} A_{total, t}}{n} \tag{3}$$

Both measures were used as a complexity measure rating, based on the findings in earlier studies where the A_{total} and A_{mean} showed to have a higher correlation with the controller's workload than the static density [24].

4. Sector complexity variables

Previous research on sector complexity showed that the aircraft intercept angle [27,28,29], speed [27] and horizontal proximity [3,16] are some of the variables that are responsible for the sector complexity. The goal of the present study is to systematically analyze the properties of the SSD due to changes in the sector design. It is hypothesized that using these properties we can obtain a more meaningful prediction of the sector's complexity (or task demand load) than existing methods.

In a first attempt, we studied the effects of aircraft streams' (that is, the airways or routes) intercept angles, the speed differences and horizontal proximity between aircraft, and also the effect of number of aircraft and their orientation on the SSD. For this purpose, several cases were studied. The cases that were being investigated involved two intercepting aircraft at variable intercept angles, route lengths, and speed vectors. Quantitative analysis was conducted on the SSD area properties for the mentioned sector variables. In the study of quantitative measurement of sector complexity, it was assumed that a denser conflict space results in a higher rating for the complexity factor. IIn later stage, a human-in-the-loop experiment will be conducted to verify the hypotheses gathered from the quantitative study and will provide a better understanding on the relationship between the SSD area properties and the workload as indicated by the subject. Figure 5 shows an example of one of the case studies with the speed vectors, route length, horizontal proximity, initial position, corresponding angle between the aircraft and the intercept angle properties. One sector complexity factor was changed at the time in order to investigate the effects of that factor on the SSD. Changes in these factors will be translated into differences in the geometry of the FBZ and area affected on the SSD.

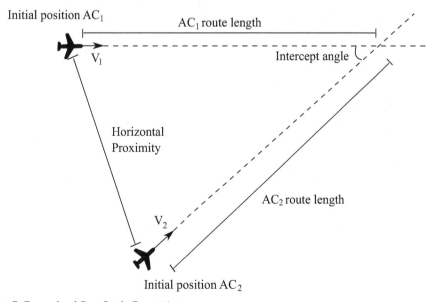

Figure 5. Example of Case Study Properties

The diagram we hereby elaborate is based on three important assumptions. First, both aircraft are on the same flight level and are not ascending or descending during the flight. Secondly, it is assumed that both aircraft have the same weight classes and will have the same minimum and maximum velocities. Lastly, the minimum separation distance, represented by a PZ with radius of 5 NM around each aircraft, is to be maintained at the same size at all time. Different complexity factors are compared using a quantitative analysis.

4.1. Horizontal proximity

Previous research on sector complexity has shown that the aircraft horizontal proximity [3,16] is one of the variables that is responsible in the sector complexity construct. There are several relationships that can be gathered from the FBZ. In order to analyze the relationship between FBZ and time to conflict and the position of aircraft, some parameters have to be determined. These parameters can be found in Figure 6 where the absolute and relative space of the FBZ was illustrated in Figure 6 (a) and (b), respectively. In the absolute space (Figure 6 (a)), two aircraft situation with distance between aircraft (d) and minimum separation distance (R) were illustrate. The FBZ is then translated into the relative space (Figure 6 (b)) where the same situation was projected with the assumption that the controlled aircraft will be in direct collision with the observed aircraft in the future. Based on the figures, it is observed that the FBZ and the corresponding Solution Space share similar geometric characteristics. These, as shown in Figure 6, make it clear that:

$$\frac{|V_{rel}|}{r} \sim \frac{|d|}{R} \tag{4}$$

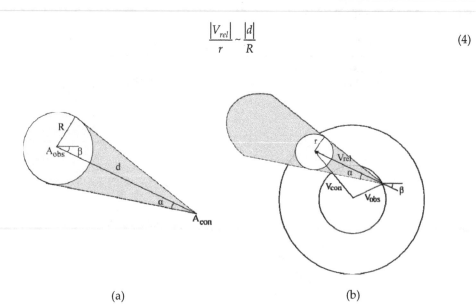

(a)   (b)

Figure 6. Projected Protected Airspace. (a) Absolute Space. (b) Relative Space.

The separation between aircraft in terms of time and horizontal proximity can be directly observed on the SSD through the width of the FBZ. A narrow FBZ translates to a longer time until loss of separation and also a larger separation distance between both aircraft. The relation can be seen in Equation (5) [34] and Equation (6), where the time (t) and distance between aircraft (d) is inversely proportional to the width (w) of the FBZ.

$$w = \frac{2R}{t \cos \alpha} \tag{5}$$

$$w = \frac{2R V_{rel}}{d \cos \alpha} \tag{6}$$

The importance of horizontal proximity has also been stressed in other research where it is indicated that aircraft that fly closer to each other have a larger weight on the Dynamic Density [3,16]. In order to see the effect of horizontal proximities on the SSD and to confirm the previous study, more than 50 position conditions with intercept angle of either 45°, 90° or 135°, were studied. To simulate horizontal proximity, aircraft were assigned with a different route length at a different time instance. It is important to ensure that only one property is changed at a time. During this study, the velocity of both aircraft was maintained at same speed at all times. The effect of the horizontal proximity on the SSD is shown in Figure 7. The situation in Figure 7 is based on aircraft flying with a fixed heading angle of 90°, while both aircraft having the same speed vector of 200 knots, but having a different route length.

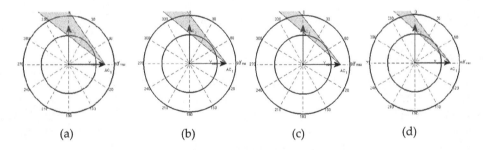

(a)   (b)   (c)   (d)

Figure 7. SSD for AC2 Observing Horizontal Proximity Changes.

From the analysis, it was found that aircraft that are further apart from each other have a narrower FBZ width than the ones being closer to each other. This can be seen in Figure 7 with aircraft progressing from being nearest (Figure 7 (a)) to furthest (Figure 7 (d)) apart from one another. The same pattern also applies to other intercept angles studied. The area affected is less dense for aircraft with a larger horizontal proximity where the area affected within the SSD decreases from 11% for the case in Figure 7 (a) to 6% for case in Figure 7 (d). This also shows that a large horizontal separation between aircraft result in a less dense SSD, thus a lower complexity metric. A narrower width also implicate that there are more options to solve a conflict. This can be seen in Figure 7, where in Figure 7 (a) and (b), there is no room for AC2 to resolve the conflict using a speed-only correction, whereas in Figure 7 (c) and (d) the conflict can be resolve by either increasing or decreasing the AC2 speed.

Similar patterns were observed with different speed settings and speed boundaries in conjunction with different intercept angles. Figure 8 illustrates the percentage area covered as a function of the horizontal distance and the intercept angle while having the same velocity vector. It can be seen from this figure that the area properties decrease with larger distances between both aircraft at any intercept angle. The regression rate of the SSD area properties against the horizontal distance is also similar with any other intercept angle as indicated by Equation (6) regarding the width of the FBZ.

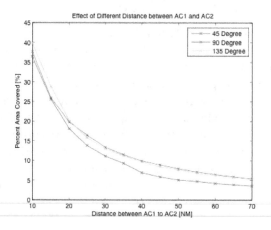

Figure 8. Percent Area Covered with Distance for Different Intercept Angle

4.2. Speed variations

A previous study by Rantenan and Nunes [27] has suggested speed as a confounding factor to conflict or intercept angles and the ability to detect a conflict. It was indicated in their research that increasing the speed differential between converging objects increased the temporal error, resulting in a lower accuracy. This is due to the fact that the controller now has to integrate two (rather than one) pieces of speed information and project their implications. This shows the importance of studying the effect of speed variations to the sector complexity, especially when coupled with the intercept angle.

A number of cases of aircraft pairs at the same distance between each other were investigated in this preliminary study. The first observation is illustrated in Figure 9 where the speed and the heading of the observed aircraft can be seen on the SSD mapping of the controlled aircraft through the position of the tip of the FBZ. This is because the FBZ is obtained by transposing the triangular-shaped conflict zone with the observed aircraft velocity vector. In a case such as seen in Figure 9 (a) to (c), an aircraft with the same horizontal separation at an intercept angle of 90° between each other will result in a different SSD as a function of the 150, 200 and 250 knots speed settings.

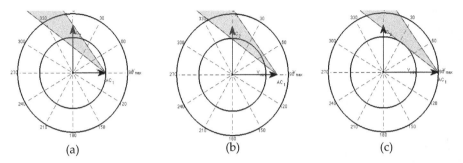

Figure 9. SSD of AC2 observing speed changes for the same aircraft position. (a) AC1 at 150 knots. (b) AC1 at 200 knots. (c) AC1 at 250 knots.

In Figure 9, AC1 will encounter a separation violation problem in the future with AC2 when the aircraft maintains its current heading and speed. However, giving speed or heading instructions to one or both aircraft can resolve the future separation issue. In this case, an increase (Figure 9 (a)) or decrease (Figure 9 (c)) in speed for AC2 will solve the future separation issue. It is not desired for on-course aircraft to change the heading angle in order to fulfill efficiency constraints, however, if required to maintain safety, it may be the proper way to resolve a conflict, such in Figure 9 (b). It is found that the higher the speed of the observed aircraft, the more the FBZ in the SSD is shifted outwards. The changes in the speed only affect the currently controlled aircraft's SSD. Because there is no change of speed for the controlled aircraft, AC2, the corresponding diagram for AC1 observing AC2 remains the same during the change of speed vector in AC1.

The total area affected on the SSD depends on the relative positions and the intercept angle of both aircraft, where a shift outwards will be translated as more or less SSD area percentage affected. This can be seen by comparing Figure 9 (a) to (c) where a shift outwards results in more area affected within the SSD, which gives the value of 8%, 11% and 15% area affected for cases (a), (b), (c), respectively. Hence it can be hypothesized that larger relative speeds can result in a higher or lower complexity metric, depending on the position and intercept angle of the aircraft.

The effect of speed differences was also investigated further for aircraft intercepting at 45°, 90° and 135° with more possible cases, and the results are illustrated in Figure 10. Differences in intercept angle, speed limit band (which may represent differences in aircraft performance limits or aircraft types) and the size of the speed limit were investigated. Figure 10 shows the effect of speed differences on a 180 - 250 knots speed band, with both AC1 and AC2 at either 30 NM or 40 NM distance from the interception point at different intercept angles. Both aircraft's initial speeds were 250 knots and to illustrate the effect of speed variations, one of the aircraft was given a gradual speed reduction toward 180 knots.

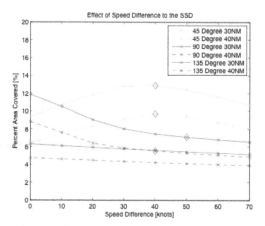

Figure 10. The SSD area values as a function of different speed settings for same aircraft position with different intercept angles.

The diamond shapes in Figure 10 indicate the minimum difference needed for aircraft not to be in a future separation violation. Based on Figure 10, the effect of speed and distance is evident with 45°, 90° and 135° intercept angles showing a decrease in the SSD area properties with a larger relative distance while maintaining the trends of the graph. In 90° and 135° cases, larger distances also indicated that a smaller speed difference (marked with diamond) was needed in order for both aircraft not to be in a future separation violation. Figure 10 also shows that aircraft flying at a smaller intercept angle needed less speed difference than aircraft flying larger intercept angle to avoid future separation violation caused by having the same flight path length to the intercept point.

The effect of the intercept angle on the other hand shows different patterns in SSD area properties in regards to the speed variations. A 45° intercept angle showed an increase of SSD area properties up until the intermediate speed limit followed by a decrease of SSD area properties with increased speed differences. However, for 90° and 135° intercept angle cases, the reduction of speed is followed by a continuing decrease in SSD area properties.

Differences in the pattern also indicated a difference in sector complexity behavior toward distinctive intercept angle. The effects of speed limit bands for 45° intercept angle cases are illustrated in Figure 11 and 12. Figure 11 (a) shows the effect of different speed band values while maintaining the same size of the controlled aircraft speed performance and Figure 12 (b) shows the effect of different sizes of the speed band. Based on both figures, irrespective of the speed band ranges (aircraft speed performance limit) or speed band size, the same pattern in area properties were found, in all eight scenarios. The only difference was the peak value of the SSD area properties (Figure 11 (a)) is greater for speed bands with higher speed limits. This is due to the fact that with the same position between both aircraft, higher speed (for AC1 in this case) indicates a higher possible relative speed (Vrel) for the maximum speed band, thus implicating a broader FBZ (can be seen in Equation (6) and Figure 11 (b)). The same pattern was illustrated with different speed band sizes (Figure 12) with higher peaks of the SSD area values for higher AC1 speeds.

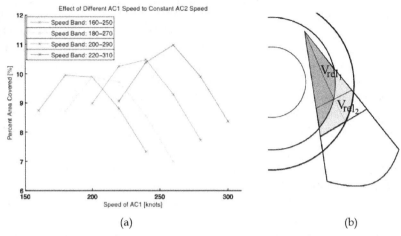

(a)   (b)

Figure 11. (a) Various speed settings for the same 45 Degree Intercept Angle with different speed limit boundaries (b) Different speed band maximum limit of the controlled aircraft

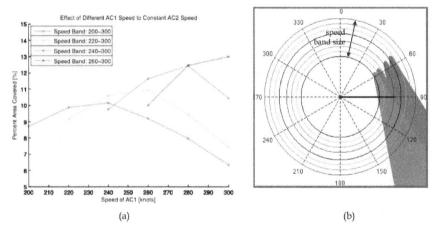

(a)   (b)

Figure 12. (a) Various speed settings for the same 45 Degree Intercept Angle with different speed band size (b) SSD of Different speed band sizes.

4.3. Intercept angle

Based on previous researches, the ability of the controller to ascertain whether or not an aircraft pair will lose separation (more commonly known as conflict detection) is affected by a variety of variables that include, but are not limited to, the convergence angle [27,28,29]. However, previous research also found that conflict angle as a factor affecting conflict detection ability, is often confounded with speed [27]. Nonetheless, in order to understand the intercept angle as part of the sector complexity measure, the effect of intercept angle on the SSD area property is important.

There are several types of crossing angles that are being studied. The main goal of the study was to investigate the effect of crossing angle towards sector complexity through the SSD. The effect of different intersection angles on the SSD is shown here for the case where the route length between AC1 and AC2 remains constant and equal at all time. Both aircraft were flying the same speed vector of 200 knots, but with different heading angles for AC2, which are 45°, 90° and 135°. The negative intercept angles were assigned for aircraft coming from the left, while positive intercept angles were assigned for aircraft coming from right. As seen here, only the changes in the heading angle were investigated, while other variables were fixed to a certain value.

From the analysis, it is found that the larger the heading angles of intersecting aircraft, the less dense the area within the SSD. Figure 13 shows the resulting SSD for different intercept angles. Figure 13 also shows the effect of aircraft coming from right (Figure 13 (a) to (c)) or from the left (Figure 13 (d) to (e)) side of the controlled aircraft. It is concluded here that aircraft coming from any direction with the same intercept angle and route length will demonstrate the same complexity measure due to the symmetrical nature of the conflict For aircraft with 45°, 90° and 135° intercept angles, the SSD area properties are 14%, 11% and 8%, respectively. The same area properties hold for the opposite angle. This also shows that a larger intercept angle results in a lower complexity metric based on the properties of the SSD, because the solution area covered with the conflict zone is smaller. However, this condition only applies if the observed aircraft has a route length larger or equal to the controlled aircraft. This also means that the condition where the effects of intercept angles on the complexity metric is only valid when the observed aircraft is approaching from a certain direction.

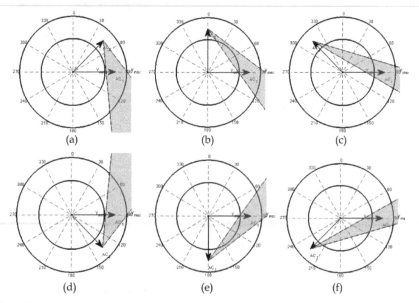

Figure 13. SSD for AC1 observing different heading angle for same aircraft speed. (a) AC2 at 45°. (b) AC2 at 90° (c) AC2 at 135° (d) AC2 at -45°. (e) AC2 at -90° (f) AC2 at -135°.

4.3.1. Front side and backside crossings

It was found that there are differences between observing an aircraft crossing in front or from the backside of the controlled aircraft with an increasing intercept angle. A case study was conducted where an aircraft observed front side and backside crossings at an angle of 45° and 135°. Both aircraft had the same speed of 220 knots and intercepted at the same point of the route, giving the same flight length for each case observed (see Figure 5). In a case where the controlled aircraft, which was farther away, was observing an intercept of an observed aircraft crossing in front at a certain angle, the area affected was increasing with an increasing intercept angle. The area affected measured in this case was 3% for 45° intercept angle (Figure 14 (a)) compared to 5% area affected for the 135° intercept angle (Figure 14 (b)). On the other hand, in a case where the controlled aircraft was observing an aircraft crossing from the backside, the area affected was decreasing with increasing intercept angle. The area affected measured in this case is 8% for 45° intercept angle (Figure 14 (c)) compared to 3% for 135° intercept angle (Figure 14 (d)). These area-affected values concluded that a slightly higher complexity metric was found with an increasing intercept angle when the observed aircraft was already present in the sector and passing the controlled aircraft from the front side. The opposite situation appeared when the observed aircraft was approaching a sector and crossed the observed aircraft from the backside.

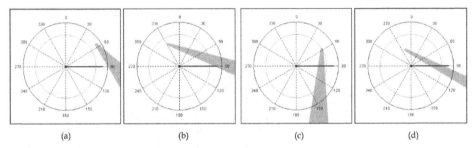

Figure 14. (a) Observed Aircraft Crossing from the front side at 45° (b) Observed Aircraft Crossing from the front side at 135° (c) Observed Aircraft Crossing from the backside at 45° (d) Observed Aircraft Crossing from the backside at 135°.

To extensively study the effect of intercept angle and the relative aircraft distance on the SSD area properties, several other cases were looked into and the results are illustrated in Figure 15. Figure 15 showed static aircraft at 35 NM distance from the intercept point, observing an incoming or a present aircraft in the sector at a variable intercept angle. Based on the initial study, it can be seen that observing present aircraft in the sector (with a distance from the intercept point less than 35 NM) will lead to an increase of SSD area properties with an increasing intercept angle. Despite this result, it was observed that a larger intercept angle for incoming aircraft (aircraft with distance more than 35 NM) results in a less dense area inside the SSD with an increasing intercept angle. The results gained here, matches the initial observations discussed earlier.

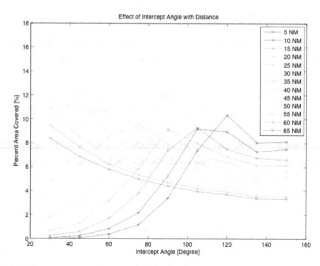

Figure 15. Plots of SSD Behavior showing the Differences in Intercept Angle and Distance to Intercept Point

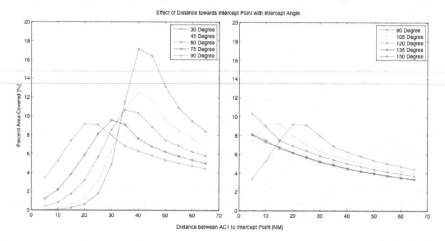

Figure 16. Plots of SSD Behavior showing the Differences in Intercept Angle and Distance to Intercept Point

Figure 16 shows the effect of intercept angle and the relative aircraft distance to the intercept point from a different perspective, where the effect of different intercept angle on the distance towards the intercept point was focused. From the figure it is observed that a larger distance for larger intercept angles (120°, 135° and 150°) results in a continuing decrease of SSD area properties, thus relating to a lower complexity metric, whereas a larger distance for smaller intercept angles (30° to 90°) result in an initial increase of SSD area properties, thus relating to a larger complexity metric and followed by decreasing SSD area properties after a certain distance (more than 35 NM). This also suggested that for a bigger intercept angle, the increase in distance always relates to a less complex situation whereas for a smaller intercept angle, the increase of distance up to a point where the length path is equal relates to a more complex situation.

4.3.2. Time to conflict

The effect of intercept angle on the sector complexity construct was also investigated from a different perspective, namely the Time to Conflict (TTC). As illustrated in Figure 17 (a), with a fixed TTC at 500 seconds, a larger conflict angle will result in lower SSD area properties, thus a lower sector complexity construct. However, this can be due to the larger distance between the aircraft for larger conflict angles, even with the same TTC value. Having said that, this also indicates that with a larger intercept angle, a later conflict detection and lower initial situation awareness are predicted. An example of the progression of a future conflict that will occur at an equal time in the future with different conflict angles is shown in Figure 17 (b). Based on Figure 17 (b), a larger conflict angle results in lower SSD area properties, and also has a faster rate of SSD progress toward total SSD occupation.

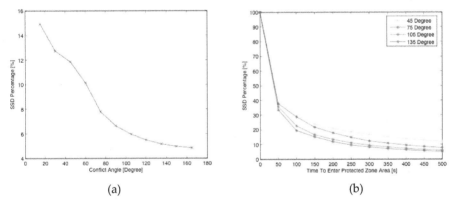

(a)   (b)

Figure 17. (a) SSD Area Properties for Different Conflict Angle Properties of Aircraft with the Same TTC. (b) SSD Area Progression with TTC for Different Conflict Angle

4.4. Number of aircraft and aircraft orientation

One of the methods to measure sector complexity is through the measurement of aircraft density. Aircraft density is one of the measures that is commonly used to have instant indication of the sector complexity. It is defined as the number of aircraft per unit of sector volume. This section discusses the effects of the number of aircraft within a sector on the SSD area properties together with the aircraft heading orientations. Figures 18 and 19 show the number of aircraft and the traffic orientation that was investigated here. An example SSD for two aircraft, AC1 and AC2 as indicated in Figure 18 and 19 were illustrated for all cases. For all four situations, all aircraft are free of conflicts. In a four-aircraft situation, illustrated in Figures 18 (a) and (d), an A_{mean} of 9% and 16%, respectively, were gathered whereas in a six-aircraft situation, illustrated in Figure 19 (a) and (d), an A_{mean} of 15% and 20%, respectively were gathered. Based on the SSD area properties, it was clear that more aircraft relates to a higher SSD area properties comparing cases in Figure 18 (a) to Figure 19 (a). The corresponding SSD also illustrates the effect of adding two aircraft to AC1 and AC2 where additional two FBZ were present in Figure 19 (b) and (c) if compared to Figure 18 (b) and (c).

This case study also agrees with the notion that aircraft orientation also influences the complexity construct of a sector through cases illustrated in Figure 18 and Figure 19. Here it can be seen that cases with converging aircraft ((Figure 18 (d) and Figure 19 (d)) result in higher SSD area properties than cases where all aircraft have an equal heading (Figure 18 (a) and Figure 19 (a)). The SSD also showed the effect of heading with Figure 18 (b) and (c) showing the FBZ of aircraft with one heading and Figure 18 (e) and (f) showing the FBZ of aircraft with several headings. The same four- aircraft situation in Figure 18 and six-aircraft situation in Figure 19 showed to be more complicated with several aircraft headings. The area properties of the situation in Figure 18 (d) (A_{mean} of 16%) and Figure 19 (a) (A_{mean} of 15%) also showed that the SSD has the potential to be a good sector complexity measure that is, it has the capability to illustrate that more aircraft does not necessarily mean higher complexity, but that the orientation of aircraft within the sector matters more.

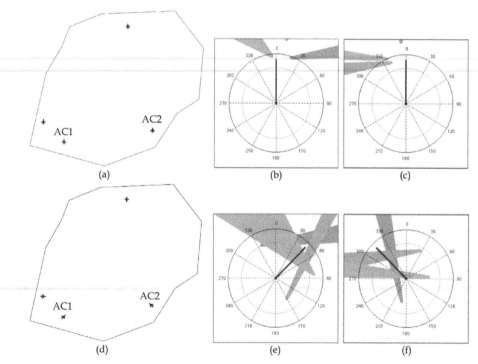

Figure 18. Different heading for same aircraft position. (a) Four Aircraft in One Heading. (b) SSD AC1. (c) SSD AC2. (d) Four Aircraft in Several Heading. (e) SSD AC1. (f) SSD AC2.

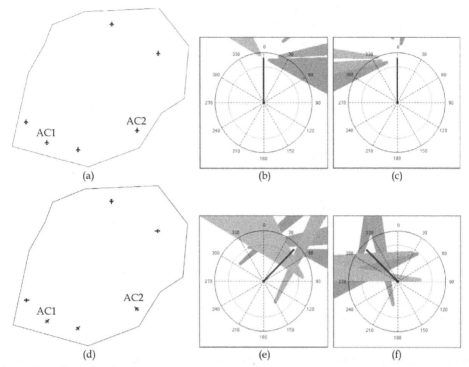

Figure 19. Different heading for same aircraft position. (a) Six Aircraft in One Heading. (b) SSD AC 2. (c) SSD AC 4. (c) Six Aircraft in Several Heading. (e) SSD AC 2. (f) SSD AC 4.

5. Solution space diagram in measuring workload

The complexity construct is an intricate topic. It is interrelated between multiple complexity variables, and altering one variable in a single scenario may result in changing other aspects of complexity. In order to measure complexity, it is hypothesized that sector complexity can be measured through the controller's workload based on the notion that the controller workload is a subjective attribute and is an effect of air traffic complexity [30]. The controller's workload can be measured based on a subjective ratings in varying scenario settings. From the many different measurement techniques for subjective workload, the Instantaneous Self Assessment (ISA) method is one of the simplest tools with which an estimate of the perceived workload can be obtained during real-time simulations or actual tasks [33]. This method requires the operator to give a rating between 1 (very low) and 5 (very high), either verbally or by means of a keyboard, of the workload he/she perceives.

While the problems encountered in Air Traffic Control have a dynamic character and workload is likely to vary over time because of the changes in the traffic situation that an ATCO is dealing with, the measurement of workload through ISA should also be made at several moments in time. To enable the SSD to become an objective sector complexity and workload measure, the correlation between the subjective ratings given by participant and the SSD area properties should be studied at several moments in time. Figure 20 shows examples of correlation study between SSD area properties and workload [24]. The plots show the subjective workload ratings in conjunction with the SSD area properties taken every minute in six different scenarios per subject. A total of 120 subjective ratings were gathered together with 120 SSD instants where SSD area assessments were conducted. With these practice, the correlation between SSD area properties and workload as indicated by controller can be evaluated.

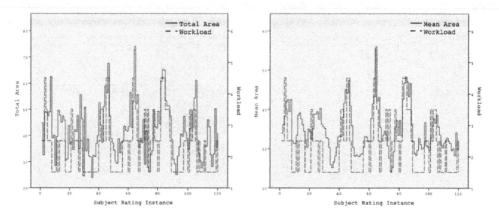

Figure 20. A_{total} and A_{mean} Plots Together with the Subjective Workload Rating as Indicated by Subject [24].

Previous experiments have shown that using the SSD area properties, a higher correlation than the static density was found [23,24]. The possibility of using the SSD in measuring workload as a function of different sector design parameters were also explored with the SSD area properties and showed to be capable of illustrating the same trend in the complexity measure with the ISA ratings [24]. However, to understand more on the complexity construct, a more focused study is needed to study different sector complexity effects on the SSD such as the number of streams, the orientation of the streams, the position of in-point and out-point of a route within the sector and etc. This preliminary study will then serve as the driver of a more elaborated research in the future.

6. Future research

The exploration of sector complexities on the Solution Space parameters and moreover workload is important in order to truly understand how workload is imposed on controllers. Because this preliminary investigation showed that various sector parameters and traffic properties are reflected by the geometry of conflict and solution spaces geometry in the SSD, the possibility of using the SSD in sector planning seems promising. This has also opened up a possibility of quantifying workload objectively using the SSD as a sector complexity and workload measure. Apart from using the SSD for offline planning purposes, having the capability to quantify sector complexity and/or workload has also a potential role in dynamic airspace assessment. This enables a more dynamic airspace sectorization or staff-planning than using the conventional maximum-number-of-aircraft limit that is primarily driven by the air traffic controller's ability to monitor and provide separation, communication and flow-control services to the aircraft in the sector.

Other than using the SSD as a sector planning aid, it is also envisioned that in the future, the SSD can be used as an operation tool. It is anticipated that by using the SSD as a display, controllers will have an additional visual assistance to navigate aircraft within the airspace. The SSD can serve as a collision avoidance tool or also a support tool for ATCOs, to indicate sector bottlenecks and hotspots.

Finally, the possibility of implementing the SSD in a three-dimensional problem is not far to reach. Initial studies have been conducted on an analytical 3D SSD [31] and an interface-based 3D SSD [32]. In the analytical solution, the 3D SSD area for the observed aircraft (A_{obs}) were comprised of two intersecting circles (both from the top and the bottom of the protected area) and the flight envelope of the controlled aircraft (A_{con}) comprising the rotation of the performance envelope around its vertical axis with 360 degrees, resulting in a donut-shaped solution space. A simplified diagram of the solution space constructed by the protected area of the observed aircraft and the flight envelope of the controlled aircraft is illustrated in Figure 21. Further studies need to be conducted to verify the capability of the 3D SSD in efficiently measuring workload or sector complexity.

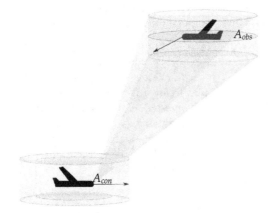

Figure 21. Two Aircraft in 3 Dimensional Conditions.

In a different study, the altitude dimensional was integrated into a 2D-based SSD ATCO display [32]. The altitude extended SSD was calculated by filtering the aircraft in accordance to their Altitude Relevance Bands and cut off the SSD conflict zones by the slowest and fastest possible climb and descent profiles. In this way, the algorithm can discard conflict zones that can never lead to a conflict. Based on this algorithm, a display prototype has been developed that is able to show the effect of altitude changes to the controller. This display will be used in the future to perform a human-in-the-loop experiment to assess the benefits of including altitude information in the 2D SSD ATCO displays.

7. Discussion

The SSD represents the spaces of velocity vectors that are conflict free. The remaining conflict areas were used as an indication of the level of difficulty that a controller has to handle. When conflict zones in the SSD occupy more area, fewer possible solutions are available to resolve future separation violations. The capability of SSD area properties in measuring the dynamic behavior of the sector was proven in previous studies [23,24]. The ongoing research is aimed at understanding the possibility of using the SSD in investigating the effects of various sector design properties on complexity and controller workload.

Based on the results gathered from the simulations, the complexity measure of intercept angle, aircraft speed, horizontal proximity, the number of aircraft, and the effect of aircraft orientation can be illustrated through the covered area percentage of the SSD. Each sector complexity factor is portrayed differently on the SSD. It is assumed that a denser area is related to a higher complexity measure. From the initial study conducted, it is concluded that a higher intercept angle, results in a smaller complexity metric, but also that this condition only applies if the observed aircraft has a route length larger or equal than the controlled aircraft. For horizontal proximity properties, it was found that further apart aircraft have a lower complexity metric. The effect of speed on the other hand depends on the position and intercept angle of the observed aircraft where a larger speed may result in higher or lower complexity metric. The number of aircraft within a sector also has a high implication on sector complexity and this was also portrayed in the SSD. However, the importance of the aircraft orientation was also an important characteristic that has an effect on the SSD area properties and thus, sector complexity.

However, it should be noted that these sector complexity parameters did not change individually at each instant, because of the dynamic behavior of the aircraft within the sector. As an initial stage of an investigation, this case study will provide the basis for hypotheses that will be tested systematically in subsequent studies. To further understand the behavior of the SSD it is important to investigate other and more combinations of sector complexity metrics. In future studies, the findings regarding the relationship between sector complexity factors and SSD metrics should be validated by means of human-in-the-loop experiments to also get the ATCO's insight on the perceived workload and how this can be related to the sector complexity mapped on the SSD.

Author details

S.M.B. Abdul Rahman, C. Borst, M. Mulder and M.M. van Paassen
Control and Simulation Division, Faculty of Aerospace Engineering,
Delft University of Technology, The Netherlands

8. References

[1] Hilburn, B. G. and Jorna, P. G. A. M., Stress (2001). Workload and Fatigue: Theory, Research and Practice., Chapter: Workload and Air Traffic Control, PA Hancock and PA Desmond (Eds.), Hillsdale, New Jersey, USA: Erlbaum, p. 384.

[2] Mogford, R. H., Guttman, J.,Morrow, S. L., and Kopardekar, P. (1995). The Complexity Construct in Air Traffic Control: A Review and Synthesis of the Literature, Tech. rep., U.S Department of Transportation, Federal Aviation Administration.

[3] Laudeman, I. V., Shelden, S., Branstron, R., and Brasil, C. (1998). Dynamic Density: An Air Traffic Management Metric, Tech. Rep. NASA-TM-1998-112226, NASA Center for AeroSpace Information.

[4] Delahaye, D. and Puechmorel, S. (2000). Air Traffic Complexity: Towards Intrinsic Metrics, 3rd USA/Europe Air Traffic Management R&D Seminar, Napoli, pp. 1–11.

[5] Airbus, S. A. S. (2010). Global Market Forecast, Technical report, Airbus.

[6] Costa, G. (1993). Evaluation of Workload in Air Traffic Controllers, Ergonomics, Vol. 36, No. 9, pp. 1111–1120.

[7] Agency, E. A. S. (2010). Annual Safety Review 2010, Technical report, European Aviation Safety Agency, Cologne, Germany.

[8] The Future of Flying (2010). Single European Sky ATM Research (SESAR) Joint Undertaking.

[9] FAA's NextGen Implementation Plan 2011 (2011). U.S Department of Transport, Federal Aviation Administration.

[10] Hilburn, B. G. (2004). Cognitive Complexity in Air Traffic Control - A Literature Review, Tech. Rep. EEC Note 04/04, EUROCONTROL, Bretigny-sur-Orge, France.

[11] Janic, M. (1997). A Model of Air Traffic Control Sector Capacity Based on Air Traffic Controller Workload, Transportation Planning and Technology, Vol. 20, pp. 311–335.

[12] Koros, A., Rocco, P. D., Panjwani, G., Ingurgio, V., and D'Arcy, J.-F. (2004). Complexity in Air Traffic Control Towers: A Field Study, Technical note dot/faa/cttn03 /14, NTIS, Springfield, Virginia.

[13] Farmer, E., & Brownson, A. (2003). Review of workload measurement, analysis and interpretation methods (No. CARE-integra-TRS-130-02-WP2).

[14] Kinney, G. C., Spahn, J., and Amato, R. A. (1977). The human element in air traffic control: Observations and analyses of the performance of controllers and supervisors in providing ATC separation services, Tech. Rep. Report No. MTR-7655, METREK Division of the MITRE Corporation, McLean, VA.

[15] Schroeder, D. J. (1982). The loss of prescribed separation between aircraft: How does it occur?, Behavioral objectives in Aviation Automated Systems Symposium, Society of Automotive Engineers, Washington, DC, pp. 257–269.

[16] Sridhar, B., Sheth, K., and Grabbe, S., (1998). Airspace Complexity and its Application in Air Traffic Management, 2nd USA/Europe Air Traffic Management R&D Seminar, Orlando, FL, pp. 1–9.

[17] Lee, K., Feron, E., and Pritchett, A. R., Describing Airspace Complexity: Airspace Response to Disturbances, Journal of Guidance, Control, and Dynamics, Vol. 32, No. 1, 2009, pp. 210–222.

[18] Hermes, P., Mulder, M., van Paassen, M. M., Beoring, J. H. L., and Huisman, H. (2009). Solution Space Based Analysis of Dificulty of Aircraft Merging Tasks, Journal of Aircraft, Vol. 46, No. 6, pp. 1–21.

[19] Koperdekar, P. and Magyarits, S. (2002). Dynamic Density: Measuring and Predicting Sector Complexity. Proceeding of the 21st Digital Avionics System Conference, Inst of Electrical and Electronics Engineers Pascataway, NJ, pp. 2C4-1-2C4-9. - 29

[20] Masalonis, A.J., Calaham, M.B. and Wanke, C.R. (2003). Dynamic Density and Complexity Metrics for Realtime Traffic Flow Management. The MITRE Corp. McLean, VA.

[21] Chatterji, G.B. & Sridhar, B. (2001). Measures for Air Traffic Controller Workload Prediction. Proceedings of the First AIAA Aircraft Technology, Integration, and Operations Forum, Los Angeles, CA. -31

[22] Laudeman, I.V., S.G. Shelden, R. Branstrom, & C.L. Brasil, (1999). Dynamic Density: An Air Traffic Management Metric. NASA-TM-1998-112226.

[23] d'Engelbronner, J., Mulder, M., van Paassen, M. M., de Stigter, S., and Huisman, H. (2010). The Use of the Dynamic Solution Space to Assess Air Traffic Controller Workload, AIAA Guidance, Navigation, and Control Conference, AIAA, Toronto, CA, p. 21, AIAA-2010-7542.

[24] Abdul Rahman S. M. B., Mulder M. and van Paassen M. M. (2011). Using the Solution Space Diagram in Measuring the Effect of Sector Complexity During Merging Scenarios, Proceeding of AIAA Guidance, Navigation, and Control Conference, Portland, Oregon.

[25] Van Dam, S. B. J. V., Abeloos, A. L. M., Mulder, M., and van Paassen, M. M. (2004). Functional Presentation of Travel Opportunities in Flexible Use Airspace: an EID of an Airborne Conflict Support Tool, IEEE International Conference on Systems, Man and Cybernatics, Vol. 1, pp. 802–808.

[26] Mercado-Velasco, G., Mulder, M., and van Paassen, M. M. (2010). Analysis of Air Traffic Controller Workload Reduction Based on the Solution Space for the Merging Task, AIAA Guidance, Navigation, and Control Conference, AIAA, Toronto, CA, p. 18, AIAA-2010-7541.

[27] Rantanen, E. M. and Nunes, A. (2005): Hierarchical Conflict Detection in Air Traffic Control, The International Journal of Aviation Psychology, 15:4, 339-362

[28] Remington, R. W., Johnston, J. C., Ruthruff, E., Gold, M., Romera. M. (2000). Visual Search in Complex Displays: Factors Affecting Conflict Detection by Air Traffic Controllers, Human Factors, Vol. 42, No. 3, Fall 2000, pg 349-366.

[29] Nunes, A. and Kirlik, A. (2005). An Empirical Study of Calibration in Air Traffic Control Expert Judgment, Proceedings of the Human Factors and Ergonomics Society 49th Annual Meeting, pp. 422-426.

[30] Koperdekar, P., Schwarzt, A., Magyarits, S. and Rhodes, J. (2009). Airspace Complexity Measurement: An Air Traffic Control Simulation Analysis. International Journal of Industrial Engineering, 16(1), pp. 61-70.

[31] Zhou, W. (2011). The 3D Solution Space: Metric to Assess Workload in Air Traffic Control. Master's Thesis. Department of Control and Simulation. Delft University of Technology

[32] Lodder, J., Comans, J., van Paassen, M. M. and Mulder, M. (2011). Altitude-extended Solution Space Diagram for Air Traffic Controllers. International Symposium on Aviation Psychology, Dayton, Ohio, USA.

[33] Tattersall, A. and Foord, P. (1996), An Experimental Evaluation of Instantaneous Self-Assessment as a Measure of Workload, Ergonomics, Vol. 39, No. 5, pp. 740–748.

[34] d'Engelbronner, J. G. (2009). Construction of a Tangent-Based Solution Space Diagram. Unpublished MSc. Thesis. Faculty of Aerospace Engineering, Delft University of Technology.

Chapter 4

Recall Performance in Air Traffic Controllers Across the 24-hr Day: Influence of Alertness and Task Demands on Recall Strategies

Claudine Mélan and Edith Galy

Additional information is available at the end of the chapter

1. Introduction

Air traffic controllers' (ATCs') work evolves constantly, concerning in particular route complexity and traffic density, but also development of supporting technology. Introducing more automation to allow more efficient ATC control and increased safety and security also requires enhanced supervisory activity, situation awareness, processing of larger amounts of data. These cognitive processes place a heavy load on ATCs' memory functions as they require item processing and recall, which are also involved in control operations such as monitoring traffic, controlling aircraft movements, managing air traffic sequences, resolving conflicts. Better understanding of memory processes and of their limitations in expert ATCs may thus be crucial for the development of future automation tools, but also for training and selection of controllers. The aim of the present contribution is to give a comprehensive overview of memorisation performance in air traffic controllers, in light of the most recent memory models. More especially, a series of experiments reveal that ATCs' memorisation performance varies in a complex manner according to both task-related factors (presentation modality, number of items, recall protocol), and task-independent factors. The latter are related more especially to shift-scheduling (time-of-day, on-shift time) and physiological capacities (alertness, automatic item processing).

2. ATCs' performance variations according to task-related factors

2.1. Information processing during control operations

En route ATC involves the processing of information relative to a variable number of aircrafts coming from different directions, at diverse speed and altitudes, and heading to various destinations or that are, on contrary, grouped in a more restricted space, thus requiring more in-depth processing in order to anticipate air-plane conflicts. Presentation modality is of interest as information about an aircraft is presented visually on a strip or script 10 to 15 min before its real-time presentation in the visual modality (radar information) or in the auditory modality (radio information). Radar information includes instantaneous level, attitude (stable, climbing, descending) and speed group, and strips present many information as aircraft call sign, aircraft type and associated speed (or power), provenance and destination, route case (estimated hours of flying), and three cases completed by ATCs with information concerning coordination information with other ATCs, or changes requested in flight.

A number of simulation studies and *in situ* ATC observations explored the way such factors may impact on memory performance. Thus, expert ATCs show high recall performance of aircrafts and their position on a sector map, and poorer recollection of details regarding an aircraft (Means et al., 1988). Higher recognition accuracy for aircrafts involved in an impending conflict compared to those that would not cross or cross in some near future confirmed the impact of aircraft status on memory performance (Gronlund et al., 1998, 2005). Simulation studies revealed higher recall when navigational messages were presented in the auditory rather than in the visual modality. In addition, performance dropped considerably when message length increased beyond three commands, while command wordiness (2 or 4 words) had only limited effects on recall (Wickens & Hollands, 2000; Barshi & Healy, 2002; Schneider et al., 2004).

2.2. Information processing according to presentation modality: The auditory superiority effect

Controlled laboratory studies have systematically explored task-related factors that may affect mnemonic performance and have led to the proposal of integrated theoretical models. More especially, encoding and processing of auditory compared to visual verbal material has systematically revealed superior recall of heard material in short-term memory (for a review, Penney, 1989). A common test in this research field contrasts silent reading and reading aloud of unrelated lists of words, nonwords, letters, or digits (see, e.g., Conrad & Hull, 1968; Frankish, 1985). It is assumed that both silent reading and reading aloud provide phonological information and that only reading aloud provides acoustic–sensory information. This effect was found for the recency part of the serial position curve in immediate free recall and in serial recall. Overt vocalization of a visually presented list by the subject produced much the same effect as did auditory presentation on the recency part of the serial position curve, but subject vocalization tended to reduce recall in the non recency part of the serial position curve. The modality effect has been observed with both written and oral recall, but seems to be slightly more marked in the former case. This auditory superiority effect, known as the modality effect, was accounted for by modality-related processing differences at different stages of the memorisation process, rather than by strategic differences.

2.2.1. Differences during processing of auditory and visually presented material

Several authors suggested that a long-lasting sensory acoustic trace would be generated for auditory presented words but not for visually presented words (Cowan, 1984; Crowder & Morton, 1969; Crowder et al., 2004; Penney, 1989). The term acoustic–sensory information refers to sensory representations of sounds. Though there is some evidence of a sensory visual trace for seen words, it would be very short-lived compared to the long-lasting sensory acoustic trace that would favour more efficient encoding of heard material. In addition, orally presented list-items would be associated with temporal cues that would not be generated by successively presented visual words. These temporal cues would then result in stronger memory traces and ensure higher recall for heard items (Frankish, 1985, 2008).

Other authors argued that the differences between the visual and the auditory information processing streams would occur when the sensory trace is processed into a short term memory trace. The working memory model (Baddeley & Hitch, 1974) includes a dedicated phonological subsystem, in which the code used to represent verbal items broadly corresponds to the phonetic level. Hence, auditory information would benefit from automatic phonological coding while visual information would require effortful phonological recoding (Baddeley, 1986, 2000; Penney, 1989). Phonological coding would favour mental rehearsal and maintenance of information in a short-term store (Baddeley, 1986), what would result in a larger number of phonologically coded memory traces in a short-term store following hearing than following seeing item-lists. According to Penney (1989) the auditory superiority would be based on the combination of a longer-lasting sensory acoustic trace and automatic phonological coding of orally presented materiel.

2.2.2. Differences during restitution of heard and seen item lists

Differences during item recall were also proposed to account for better immediate recall of heard than of seen materiel. This research is based in particular on the robust finding that when participants recall a sequence of spoken digits the last one is almost always correctly recalled, but if the same sequence is presented visually, recall of the final item is relatively poor. According to Crowder and Morton (1969), this auditory recency effect indicates that an acoustically coded representation of the final list item is maintained in a sensory store, while representations of earlier items are overwritten by successively incoming speech sounds.

Alternatively, Cowan and co-workers accounted for the recency effect by suggesting greater resistance to output interference for heard compared to seen verbal material (Cowan et al., 2002; Harveay & Beaman, 2007; Madigan, 1971). Output interference is defined as the degradation of memory representations as recall proceeds across output positions. Evidence in favour of a greater resistance to interference for heard items came from studies reporting a marked auditory superiority when restitution was based on a free recall procedure (without providing any cues) and a reduced effect when a recognition procedure was used (list items are presented together with new items). Item recognition may be explained by an impression of familiarity for the list-items and would thus not require in-depth processing.

In contrast, free recall requires subjects to explicitly recall each item, thereby degrading the remaining memory traces (Brébion et al., 2005).

Lower output interference has also been proposed to account for the auditory advantage of sentence recall (Rummer & Schweppe, 2005). In line with this hypothesis, Beaman and Morton (2000) showed that following presentation of 16-item lists, subjects preferentially recalled 2-, 3- and 4-item sub-sequences (of the form 15-16, 14-15-16, 13-14-15-16) from the end of the lists. However, while end sub-sequences were recalled with a similar frequency in both modalities on the opening run of a trial, they were recalled significantly more often for heard than for seen items during the course of a trial.

2.3. Auditory superiority for heard versus seen item-lists in ATCs

In light of these findings we explored whether output interference while recalling seen items may be reduced in participants well-practiced in processing visual information, i.e. ATCs habitually processing visual information on a radar. In addition we tested to what extent the effect may be explained by memory load, typically explored by presenting a list of items of variable length. Previous research with this kind of procedure has shown that the amount of proactive interference is much less for smaller set sizes (Oberauer & Vockenberg, 2009) although there can be some proactive interference even at small set sizes (Carroll et al., 2010). We used the procedure described by Beaman and Morton (2000) to explore free recall of heard versus seen 6- and 9-item lists in 15 volunteer ATCs of an en-route centre in southern France. Participants were aged 31.3 years (range: 27 to 42 years old) and had been working for 7 years and 4 months (range: 3 to18 years) in the control centre. The sub-sequences recalled by ATCs at initial and non-initial output positions are summarized in table 1. Analyses revealed significant higher mean numbers of items recalled following hearing (4.9 and 3.6 respectively for 6- and 9- item lists) than seeing the lists (4.2 and 3.2 respectively).

	Visual	Auditory
Terminal item sequences	6-word lists	
6	13	7
5, 6	13	17
4, 5, 6	12	12
3, 4, 5, 6	3	2
2, 3, 4, 5, 6	7	2
1, 2, 3, 4, 5, 6	10	27
Terminal item sequences	9-word lists	
9	24	20
8, 9	22	18
7, 8, 9	7	14
6, 7, 8, 9	3	4
5, 6, 7, 8, 9	0	3
4, 5, 6, 7, 8, 9	1	1

Recall Performance in Air Traffic Controllers Across the 24-hr Day: Influence of Alertness and Task Demands on Recall Strategies

53

	Visual	Auditory
Terminal item sequences	6-word lists	
5, 6	16	34
4, 5, 6	3	13
3, 4, 5, 6	1	2
Terminal item sequences	9-word lists	
8, 9	14	30
7, 8, 9	1	4
6, 7, 8, 9	1	2

Table 1. Number of each of the terminal item sequences as the opening run (upper panel) and in non-initial response positions (lower panel) during free recall of 6- and 9-word lists with auditory and visual presentation (Galy et al., 2010).

As shown in table 1, ATCs recalled complete 6-item lists three times more frequently following auditory (27) than following visual list (10) presentation, while recall of five-item sub-sequences (2-3-4-5-6) was rare in both modalities (2 vs. 7 respectively). These results further indicate that occurrences for heard lists were tenfold more frequent for complete six-item lists than for five-item sub-sequences, while no such difference occurred for seen lists. Occurrences of equivalent end sub-sequences of 9-item lists (5-6-7-8-9; 4-5-6-7-8-9) ranged between zero and three in both modalities. The auditory recall advantage of these lists appeared to result from higher occurrences of ordered 2- and 3-item end sub-sequences in *other than initial recall positions.*

Taken together, the findings show that ATCs would spontaneously adopt the output strategy consisting of uttering end sub-sequences more frequently for heard than for seen items, leading to, or contributing to significant higher overall performance. They thus further stress the proposal that "auditory presentation seems to protect the end of the list from output interference" (Cowan et al., 2002, p.168). In favour of this hypothesis, these authors showed that the auditory advantage is even more pronounced when output interference is high. Alternatively it has been proposed that the auditory advantage that extends over the last few serial positions is retrieved independently for each item from an echoic trace (Frankish, 2008). According to this author, pronounced recency in immediate serial recall is limited to stimuli that engage the perceptual mechanism involved in linguistic decoding of speech. Further research intended to disentangle different sources of sensory information at input. Thus, Rummer and Schweppe (2005) observed a modality effect for spoken sentences compared to conditions without acoustic–sensory information, i.e., both silent reading and mouthing. As the latter two conditions did not differ from each other, the results would rule out articulatory information at input as a source of the modality effect. Differences due to output modality were also investigated. For written recall the auditory advantage was larger with high than with low output interference, while this difference was not maintained for spoken recall (Harvey & Beaman, 2007). Taken together, the data suggest that both superior auditory encoding and reduced output interference would contribute to the auditory modality effect.

As ATCs are well-practiced in processing successively presented visual information, on contrary to participants in the above-cited laboratory experiments, the present findings favour the idea that differential physiological features may characterize the visual and auditory information processing streams (Penney, 1989). The modality effect may then be implemented in the current models of spoken and of written words, which are divided on the relation between the different levels of processing, which will not be discussed in this issue. Briefly, in some models the transmission of information through successive levels of representation is represented as a unidirectional flow within a feed forward network. Perceptual analysis begins with encoding of acoustic features, which are then translated into phonetic and then lexical representations. In contrast, interactive activation models propose that communication between these levels is bidirectional. Whenever a lexical unit becomes active, feedback connections boost activation of the units that represent its constituent phonemes (for a review, Frankish, 2008).

3. Shift-scheduling and ATCs' functional state

Like in other safety-related job situations, ATC requires operators to work successive shifts; i.e. different teams work in succession to cover the whole 24h-day. As a consequence, controllers are subjected to the negative impact of shift work on biological rhythms, sleep, job performance, and psychological measures (Costa, 1999; Della Rocco & Nesthus, 2005; Dinges et al., 1997; Folkard & Tucker, 2003).

3.1. Regulation mechanisms of circadian variations

It is now largely accepted that the negative effects of shift work result from a disruption of the habitual circadian regulation of physiological and psychological measures (Costa, 2003; Siegrist, 2010). Circadian variations across the 24h-day are under control of two endogenous systems, the homeostatic system expressed by the fatigue accumulating since awakening, and the circadian system evidenced by a sinusoidal variation across the 24h period. Hence, shift work is systematically associated with a cumulative sleep deficit (homeostatic system), and a decreased amplitude of sinusoidal variations (circadian system), thereby interfering with the two powerful factors limiting human ability and, as a consequence, safety and security (Akerstedt, 1991; Akerstedt et al., 2004; Dinges et al., 1997; Folkard & Akerstedt, 1992; Tucker et al., 2006). As stated by Akerstedt (2007, p. 209) "Being exposed to the circadian low (during work/activity), extended time awake or reduced duration of sleep will impair performance".

3.2. Circadian and non circadian variations of subjective and physiological measures in shift-workers

In order to investigate a person's functional state, behavioural, physiological or subjective measures have been used. It is generally accepted that these measures are subjected to a circadian rhythmicity. Shift work effects have mostly been documented by reporting decreased self-rated alertness (Akerstedt & Gillberg, 1990; Galy et al., 2008) and increased

occurrences of incidents on the night shift when the circadian decline in human capabilities
is further aggravated by a chronic sleep deficit and fatigue (Costa, 2003; Folkard & Akersted,
2003). Laboratory studies have established that alertness is low in the morning, increases
during the day until the late afternoon, before decreasing in the evening and reach a
minimal level on early morning hours. The shift work literature revealed that in real-job
situations alertness also varies with time of day and that the typical diurnal trend would
only be marginally modified by shift work scheduling features. However, early morning
shifts, extended shift duration and repeated night shifts have been shown to be associated
with increased sleepiness, more especially during the last half of extended shifts,
particularly on night shifts (Kecklund et al., 1997; Rosa, 1995), but also on day shifts (Tucker
et al., 1998).

Circadian variations appear to be less consistent for other psychological measures recorded
in several shift work studies, and in particular for self-reported tension (Folkard, 1990;
Kecklund et al., 1997; Monk et al., 1985; Owens et al., 2000 ; Prizmic et al., 1995). While some
studies reported a circadian trend for perceived tension, others did not, and still others
reported an atypical trend. More especially, operators supervising a satellite across 24h-day
displayed significant increased self-rated tension and heart rate on the first hour of each
shift, even on the night-shift, despite a lower baseline level for heart rate during the night
(Cariou et al., 2008). In contrast, when the same satellite controllers rated Thayer's (1989)
Activation-Deactivation checklist, their alertness level was highly correlated with their body
temperature (Fig. 1), largely considered as an index of subjects' functional state. Both
measures followed a typical circadian trend, indicating a strong dependency of these
measures on the endogenous regulation systems.

Taken together with the shift-work literature, these results indicate that some subjective
and objective measures (here alertness and body temperature) show a strong dependency
on the endogenous regulation systems, as they display a circadian trend in different shift-
work conditions, like in controlled laboratory conditions, despite minor variations in this
trend by external factors. This then implies that working during the circadian trough
requires an additional effort as operators' functional state is at its lowest level. Most
interestingly, other measures (here subjective tension and heart rate), which are known to
display a circadian trend in controlled laboratory conditions, are much more influenced
by external factors, so that the trends may vary considerably between situations (or job-
situations). Andorre and Queinnec (1998) reported a significant increase on the first shift-
hour for real-job performance (pages checked on a computer-screen) in operators
controlling a chemical production process. In both studies the atypical trend was
interpreted as indicating enhanced cognitive demands following shift take-over in job-
situations concerned with supervisory control of a dynamic process. Thus, in shift-work
conditions, some psychological and physiological measures, and in particular those that
are and other stress-sensitive, would be largely influenced by environmental factors,
including meal-timing, task demands, time-pressure and so on, which may mask the
otherwise circadian trend of these measures (Averty et al., 2004; Brookings et al., 1996;
Khaleque, 1984; Rose et al., 1982).

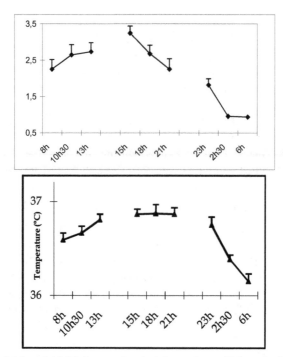

Figure 1. Upper panel: Mean (+/-S.E.) alertness level on 3 occasions within each of three shifts (1h following shift-start, middle, and 1h before shift-end). Lower panel: Mean (+/-S.E.) sublingual temperature on the same recordings (Cariou et al., 2008).

3.3. Circadian variations of subjective measures in ATCs

In light of the inconsistencies of the findings in the literature concerning other subjective measures than alertness, we investigated whether ATCs displayed typical circadian trends for subjective measures and whether on-shift time would modulate these measures. In this job-situation, traffic density variations across the 24h-day determine, partly at least, shift schedules that include in particular overlapping shifts and variable shift-duration.

Shift	Time of recording			
	01:00	07:00	13:00	19:00
06:30 -14:00		0h30	6h30	
07:00 - 17:30		0h00	6h00	
09:00 - 20:00			4h00	10h00
11:00 - 20:00			2h00	8h00
15:30 - 23:00				3h30
20:00 - 07:00	5h00			

Table 2. Time on duty of controllers on each of six shifts at the time each recording was performed (Mélan et al., 2007).

Therefore, 15 volunteer ATCs were asked to rate Thayer's (1989) checklist on 01:00, 07:00, 13:00, 19:00 by indicating whether they were on shift for four hours at most or for six hours at least (table 2). Statistical analyses revealed significant time-of-day and time-on-duty effects for both measures. ATCs rated alertness at a lower level at 01:00 and 07:00 than at 13:00 and 19:00 and tension at 07:00 compared to 01:00, 13:00 and 19:00. However, while self-rated tension was higher following long on-shift time, the opposite pattern was observed for self-rated alertness. In other words, when controllers started day-duty they experienced high alertness and low tension, whereas they reported decreased alertness and higher subjective tension after several hours on duty.

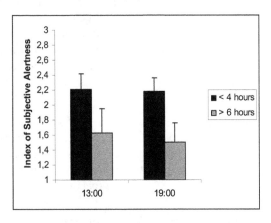

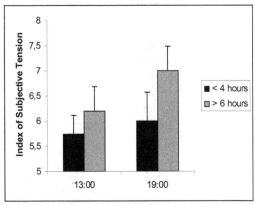

Figure 2. Mean (+/-SE) subjective alertness (upper pannel) and tension (lower pannel) in ATCs as a function of recording time (Mélan et al., 2007).

These data favour the interpretation that lower alertness reported by 12-h workers compared to 8-h workers on the early afternoon resulted from the fact that the former were on the second half of duty while the latter started their afternoon-duty (Tucker et al., 1998). Reduced day time alertness was observed on shifts starting late in the morning, though controllers had probably sufficient sleep on the night prior the shift. This raises the possibility that the chronic sleep deficit observed in shift-workers may, partly, account for decreased day-time alertness.

Further, both measures were negatively correlated, indicating that the lower ATCs quoted alertness, the higher they quoted tension. The data thus extend the findings of a circadian variation of subjective measures in ATCs and they favour the interpretation that in stress-related job-situations enhanced tension may compensate for decreased alertness (Thayer, 1989), thereby enabling the maintenance of safety. Further investigations will however be necessary to establish firmly whether ATCs' tension-ratings indicate a direct influence by environmental factors (i.e. hours at work, heavy traffic), or whether the observed variations are merely the consequence of their functional state, as indicated by the negative correlation between tension and alertness. Thus, air traffic control activities differ indeed between day- and night-work, as high traffic on day-time involves sustained periods of high task requirements and attentional demands, whereas low traffic during the night would favor boredom proneness and sleepiness (Costa, 1999; Lille & Cheliout, 1982; Luna et al., 1997).

In recent years, most studies exploring the impact of shift work on health and well-being have reported troubles in psychological and social well-being, performance efficiency and increased stress levels (Costa, 2003). Some of these issues will be highlighted in the next section by exploring the relationship between ATCs' physiological state and their performance efficiency during short-term recall of verbal material.

4. Complex interactions between task-related factors and ATCs' functional state

The shift-work literature shows that psychological measures differ not only according to time of day but also according to task characteristics, with high performance for tasks requiring rather automatic processing such as short-term memory task when alertness is low (in the morning), and high performance in more demanding tasks, relying for instance on long-term memory processing when alertness is high (in the evening). Laboratory studies reported indeed higher immediate recall in short-term memory tasks in the morning and enhanced delayed recall from long-term memory in the evening (Folkard, 1979; Folkard et al., 1976; Folkard & Monk, 1980; Monk & Embrey, 1981). Further, recall performance was reported to be higher in the morning when a recognition procedure was used, and in the afternoon when participants had to recall the text with a more demanding free recall procedure (Lorenzetti & Natale, 1996; Oakhill & Davies, 1989). Free recall would be cognitively more demanding as it involves an active search of memory traces in the absence of any cues, while item recognition would rely on less demanding item familiarity (Brébion et al., 2005; Mandler, 1980; Prince et al., 2005).

In light of the findings of the literature, it was important to investigate whether and to what extent the above-reported modality effect may be sensitive to alertness variations, but also to the cognitive effort required to remember the verbal material (Mélan, et al., 2007). ATCs' recall performance was recorded on different times of the day (01:00, 07:00, 13:00 and 19:00)

Recall Performance in Air Traffic Controllers Across the 24-hr Day: Influence of Alertness and Task Demands
on Recall Strategies

59

while varying item modality (auditory or visual) during encoding and restitution, list-length
(6- and 9-item lists) and restitution processes (recognition and free recall).

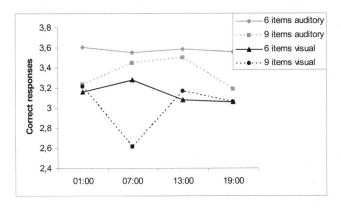

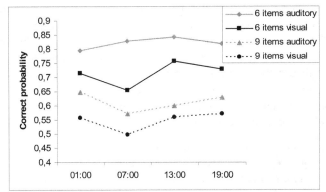

Figure 3. Mean number of correct responses in a probe recognition task (upper panel) and in a free
recall task (lower panel) on each of four recordings according to the number of items presented (6- vs. 9-
item lists) and presentation modality (auditory vs. visual; Mélan, et al., 2007).

Overall, ATCs' performance was lowest on 07:00 compared to 13:00 and 19:00, in particular
for visual items, for the longer lists and in the free recall task (Fig.3). In consequence, when
the tasks involved more demanding processing performance was decreased particularly
when alertness was low. Interestingly, significant main effects occurred for time-of-day, list-
length and modality with the free recall procedure, while the three variables interacted
when restitution was based on a recognition procedure. In that case, performance dropped
in the early morning only for visual 9-item lists, i.e. in the more demanding task conditions.

In the literature, differential time-of-day effects on participants' recall performance in
recognition and free recall tasks were accounted for by similar processing differences
(Folkard, 1979; Folkard, et al., 1976; Folkard & Monk, 1980; Lorenzetti & Natale, 1996;
Oakhill & Davies, 1989). Accordingly, the finding of an overall effect on free recall but not
on recognition performance may indicate that external factors (time of day) more readily

impacted on ATCs' task performance when deeper processing was required to solve a task. As indicated above, item recognition requires less in-depth processing of the to-be-remembered material than free recall (Brébion et al., 2005; Mandler, 1980; Prince et al., 2005). Thus, task-dependent factors (modality, list-length and recall procedure in the present case) even further impact on ATCs' performance when their functional state is low.

5. Discussion

The main finding of the present contribution is that ATCs' performance depends in a complex manner from the task to be performed, but also from job organisation (shift schedules, shift-duration) and from physiological aspects (alertness, sensory and cognitive processing). Even though it is difficult to generalize between different job-situations, given in particular differences between activities, to-be-performed tasks and shift-scheduling, the data confirm findings reported in other safety-related job-situations for security agents in a nuclear power plant and for operators controlling satellites (Costa, 1999; Cariou et al., 2008; Galy et al., 2008). Working during the night causes a mismatch between the endogenous circadian timing system and the environmental synchronizers (the light/dark cycle in particular), with consequent disturbances of the normal circadian rhythms of psycho-physiological functions, beginning with the sleep/wake rhythm, and thus operators' alertness. In addition to the disruptive effects of shift work on performance efficiency, its impact on health and well-being are now well-documented (Costa, 2003; Siegrist, 2010). In this respect, some international directives have recently stressed the need for the careful organization of shift and night work and the protection of shift workers' health.

Elsewhere, the present findings favour the idea of a more general model to account for the complex interactions reported so far. In this regard, we recently extended Sweller's cognitive load theory elaborated in the educational field (Sweller, 1988; 1994), to a real-job situation (Galy et al., 2012).

5.1. Towards an integrated model of mental load during ATC?

Sweller defined three categories of cognitive load in order to embrace the complexities of a given task, the conditions in which it is performed, and subject-related variables. "Intrinsic cognitive load" refers to the load induced by the material to be processed, such as task difficulty that is defined in particular by the number of items to be processed, and by item interactivity (Ayres, 2006; Kalyuga et al., 2003; Sweller & Chandler, 1994). "Extraneous mental workload" refers to the load induced by external factors, including work situation, work organization, time pressure, and noise (Sweller, 1994). Finally, "germane mental workload" corresponds to the load induced by conscious application of strategies to solve tasks more easily (Schnotz & Kürschner, 2007). We showed an additive effect of intrinsic load (task difficulty) and extrinsic load (time pressure) factors, and that this effect was only observed in the morning. In other words, when participants performed a difficult task under high time pressure in the morning, when alertness and thus available mental resources were low, they probably had to use specific strategies generating germane load. This sequence of events would have led to decreased performance in the morning, while no such effect was observed in the afternoon, when alertness and thus available mental resources were high.

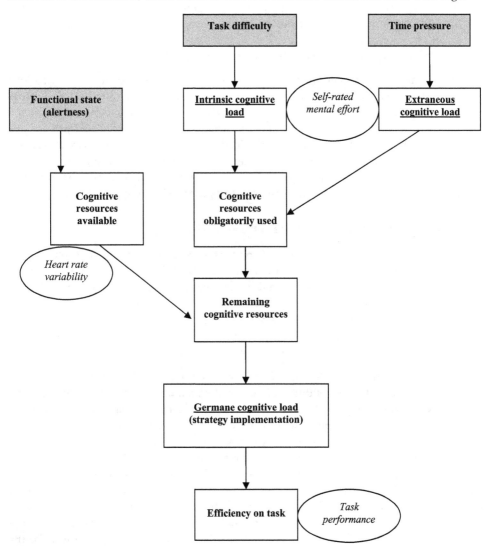

Figure 4. Graphic representation of putative relationships between cognitive load factors and cognitive load categories (Galy et al., 2012).

This model, summarized in figure 4, further indicates that the different cognitive load measures used in the study, i.e. subjective measures (self-rated effort), behavioural measures (correct responses) and psychophysiological measures (heart rate variability) display a differential sensitivity to the three kinds of load factors investigated. More especially, heart rate variability increased with germane load, operationalized by high self-reported alertness, whereas participants' self-rated mental effort was sensitive to both extrinsic and intrinsic load and task performance was determined both by alertness and by an interaction between task difficulty and time pressure. The latter results stress the difficulties encountered in order to identify a reliable measure of cognitive load on one hand (Backs & Seljos, 1994; Brookings et al., 1996; Carroll et al., 1986), and suggest that it may be hazardous to generalize the proposed model to other situations, on the other hand.

With these limitations in mind, it may nevertheless be challenging to test the mental load model in ATCs, given that the findings reported in the previous sections indicate how work organization (shift schedules, shift-duration) affected alertness (germane mental load), and that when ATCs' alertness was low, their mnemonic performance was lower in the demanding task conditions (intrinsic mental load). Further, as conflicts between aircrafts require ATCs to make the right decisions under high time pressure and to give ground-to-air instructions in a limited time, it seems indeed worthwhile to include this extrinsic load factor to a more integrative approach of mental load generated during ATC activities.

5.2. Experimental designs to explore ATC activities

Experimental paradigms like the ones used in the studies reported in this contribution may be regarded as simplified models of real-life activities. As such, the results reported may be of interest for ATC activities involving similar cognitive processes than those explored in the experimental paradigms. This is in particular the case for task-related factors, as the influence of such factors may only be demonstrated in controlled study conditions. In this regard, experimental approaches are complementary to *in situ* observations, which point to the relevant aspects to be explored more systematically by using experimental designs.

Further, the observations reported here were performed while ATCs were in their habitual work environment and work conditions. In contrast, in a number of field studies subjective data across the 24h-day are collected retrospectively, i.e. participants are asked to rate these measures on a single session by remembering what they perceived for instance during the night or during the morning shift. In contrast, subjective and performance measures reported in the present contribution were collected in real-time fashion, in order to gain some insight into ATCs' cognitive abilities in real-job conditions. This is particularly important when investigating job-activities performed on a continuous 24h-day, like ATC. The data reported here clearly show that shift-work features (time-of-day and time-on-shift) are indeed critical factors that impact not only on operators' functional state, but also on a number of psychological measures. ATCs' information processing ability is crucial for the safety of the air traffic management system as well as for the sector capacity of a given complexity in a particular time driving the overall system performances.

The present findings may be relevant to ensure productivity and/or safety in job-situations involving supervisory control, and in particular ATC, all relying on processing visual and auditory information from various sources on control panels and interfaces. The present findings suggest that short sequences of auditory information would less readily tax controller's processing capacities than longer sequences and/or visual material. As stated by

Frankish (2008), all models of speech perception incorporate some form of auditory short term memory, because speech comprehension requires the integration of information from successive elements. The findings reported here may thus be useful for various control operations and in particular during conflict resolution, when controllers' memory span may be more readily taxed. As shown by several studies, short-term recall of navigational messages decreases when message length increases beyond three commands (Wickens & Hollands, 2000; Barshi & Healy, 2002; Schneider et al., 2004). Taken together with several other studies, the present results may also be useful for ATC selection and training, as they stress the importance of using tests that manipulate, in addition to the more traditional quantitative aspects of memory, more qualitative parameters, such as presentation modality or information type and aircraft status (Gronlund et al., 2005; Means et al., 1988).

ATC, like several other safety-related job-activities rely preferentially on visual interfaces, most probably because graphical representations enhance the understanding of complex principles or spatial relations, for instance flight direction, speed and other information that an operator needs to synthetize in order to solve conflicting situations for instance. Within the auditory stream, successive items are strongly associated; in contrast, in the visual modality, it is simultaneously presented items that are strongly associated (Penney, 1989). Speech, with its linear progression through time does not bear these properties. In addition, sound and speech generate considerable noise when compared to silent reading of messages or integration of graphical representations.

However, for precisely the same reason, i.e. the disturbance induced by auditory information, this modality is typically used for alarm devices, as one cannot avoid an auditory signal. Indeed, the physical nature of sounds, i.e. waves that cross space in all directions, ensures that this kind of information automatically reaches and stimulates the auditory receptors and is transferred thereafter to the auditory cortex. There is thus a pretty good chance for heard messages or signals to be processed, even without paying attention to one's environment or when a subject's motivation or alertness are decreased for some reason. Furthermore, in addition to a specific activation of the auditory system, sounds also induce a non-specific activation of the nervous system, both the autonomous nervous system as evidenced for instance by increased heart-rate, and the central nervous system, as indicated by enhanced arousal. On contrary, written messages may be processed by the visual system only once a subject has oriented his/her gaze on the specific location where the message has occurred. Thus, unless the operator turns his/her attention to a particular location on a radar screen, he/she will not be able to appreciate whether for instance some airplane may be in a difficult position. Research concerning participant's focus of attention that would reflect conscious awareness and its relation to visual working memory tasks have been reviewed recently (Cowan, 2011).

In light of these considerations it appears that supervisory control greatly benefits from a well-weighted combination of visual representations of complex multi-dimensional data and auditory presentation of essential information which have to be maintained in a short-term memory for problem resolution.

Author details

Claudine Mélan
Shift-work and Cognition Laboratory, Toulouse University, France

Edith Galy
Research Centre in the Psychology of Cognition, Language, and Emotion, Aix-Marseille University, France

6. References

Akerstedt, T. (1991). Sleepiness at work: Effect of irregular work hours. In: *Sleep, sleepiness and performance*, T. Monk, (Ed.), 129-152, John Wiley & Sons ltd,

Akerstedt, T. (2007). Altered sleep/wake patterns and mental performance. *Physiology and Behavior*, Vol.90, No.2-3, pp. 209-18

Akerstedt T, Folkard S. & Fortin C. (2004). Predictions from the three-process model of alertness. *Aviation Space and Environmental Medicine*, Vol.75, pp.75-83

Akerstedt, T. & Gillberg, M. (1990). Subjective and objective sleepiness in the active individual. *International Journal of Neuroscience*, Vol.52, pp. 29-37

Andorre, V. & Quéinnec, Y. (1998). Changes in supervisory activity of a continuous process during night and day shifts. *International Journal of Industrial Ergonomics*, Vol.21, pp. 179-186

Averty, P., Collet, C., Dittmar, A., Athenes, S. & Vernet-Maury E. (2004). Mental workload in air traffic control: an index constructed from field tests. Aviation Space and Environmental Medicine, Vol.75, pp. 333-341

Ayres, P. (2006). Using subjective measures to detect variations of intrinsic cognitive load within problems. *Learning and Instruction*, Vol.16, pp. 389–400

Backs, R.W. & Seljos, K.A. (1994). Metabolic and cardio-respiratory of mental effort: the effects of level of difficulty in a working memory task. *International Journal of Psychophysiology*, Vol.16, pp. 57-69

Baddeley, A. D. (1986). *Working memory*. Oxford: Oxford University Press, UK

Baddeley, A. D. (2000). The episodic buffer: a new component of working memory? *Trends in Cognitive Science*, Vol.4, No.11, pp. 417–423

Baddeley, A. D. & Hitch, G. (1974). Working memory. In: *The Psychology of Learning and Motivation*, G.A. Bower, (Ed.), 48–79, Academic Press

Brookings, J.B, Wilson, G.F. & Swain, C.R. (1996). Psychophysiological responses to changes in workload during simulated air traffic control. *Biological Psychology*, Vol.42, pp. 361-377

Beaman, C.P. & Morton, J. (2000). The separate but related origins of the recency effect and the modality effect in free recall. *Cognition*, Vol.77, 59-65

Barshi, I. & Healy, A. F. (2002). The effects of mental representation on performance in a navigation task. *Memory and Cognition*, Vol.30, pp. 1189-1203

Brébion, G., David, A.S., Bressan, R.A. & Pilowski, L.S. (2005). Word frequency effects on free recall and recognition in patients with schizophrenia. *Journal of Psychiatry Research*, Vol.39, pp. 215-222

Conrad, R. & Hull, A.J. (1968) Input modality and the serial position curve in short-term memory. *Psychonomic Science*, Vol.10, no.4), pp.135-136.

Cariou, M.; Galy, E. & Mélan, C. (2008). Differential 24-h variations of alertness and subjective tension in process controllers: investigation of a relationship with body temperature and heart rate. *Chronobiology International*, Vol.25, No.4, pp. 97-609

Carroll, D., Turner J.R. & Hellawell, J.C. (1986). Heart rate and oxygen consumption during active psychological challenge: the effects of level of difficulty. Psychophysiology. 23:174-181

Carroll, L.M.; Jalbert, A., Penney, A.M., Neath, I., Surprenant, A.M. & Tehan, G. (2010). Evidence for proactive interference in the focus of attention of working memory. *Canadian Journal of Experimental Psychology*, Vol.64, pp. 208–214

Costa, G. (1996). The impact of shift and nightwork on health. *Applied Ergonomics*, Vol.27, no.1, pp. 9-16

Costa, G. (1999). Fatigue and biological rhythms. In: *Handbook of aviation human factors*. D.J. Garland, J.A. Wise & V.D. Hopkin, (Eds.), 235-255, London: Erlbaum

Costa, G. (2003). Shift work and occupational medicine: An overview. *Occupational Medicine*, Vol.53, pp. 83-88

Cowan, N. (1984). On short and long auditory stores. *Psychological Bulletin*, Vol.96, pp. 341-370

Cowan, N. (2011).The focus of attention as observed in visual working memory tasks: Making sense of competing claims. *Neuropsychologia*, Vol.49, pp. 1401–1406

Cowan, N.; Saults, J.S., Elliott, E.M. & Moreno, M.V. (2002). Deconfounding serial recall. *Journal of Memory and Language*, Vol.46, pp. 153-177

Cowan, N.; Saults, J.S. & Brown, G. D. (2004). On the auditory modality superiority effect in serial recall: separating input and output factors. *Journal of Memory and Language*, Vol.46, pp. 153-177

Crowder, R.G. & Morton, J. (1969). Precategorical acoustic storage (PAS). *Perception and Psychophysics*, Vol.5, pp. 365-373

Frankish, C. (1985). Modality-specific grouping effects in short-term memory. *Journal of Memory and Language*, Vol.24, pp. 200-209

Frankish, C. (2008). Precategorical acoustic storage and the perception of speech. *Journal of Memory and Language*, Vol.58, pp. 815–836

Della Rocco, P.S. & Nesthus, T.E. (2005). Shift-work and air traffic control: Transitioning research results to the workforce. In: *Human factors impacts in air traffic management*. B. Kirwan, M. Rogers, D. Schäfer, (Eds), 243-278, Aldershot, UK: Ashgate

Dinges, L.S.; Pack, F., Williams, W., Gillen, K.A., Powell, J.W., Ott, G.E., Aptowicz, C. & Pack, A.I. (1997). Cumulative sleepiness, mood disturbance, and psychomotor vigilance performance decrements during a week of sleep restricted to 4-5 hours per night. *Sleep*, Vol.20, No.4, pp. 267-277

Folkard, S. (1979). Times of day and level of processing. *Memory and Cognition*, Vol.7, PP. 247-252.

Folkard, S.; Knauth, P., Monk, T.H. & Rutenfranz, J. (1976). The effect of memory load on the circadian variation in performance efficiency under a rapidly rotating shift system. *Ergonomics*, Vol.19, pp. 479-488.

Folkard, S. & Monk, T. H. (1980). Circadian rhythms in human memory. *British Journal of Psychology*, Vol.71, pp. 295-307.

Folkard, S. & Akerstedt, T. (1992). A three-process model of the regulation of alertness-sleepiness. In: *Sleep, arousal, and performance*, R.J. Broughton & R.D. Ogilvie, (Eds.), 11-26, Boston: Birkhaüser

Folkard, S. & Tucker, P. (2003). Shift work, safety and productivity. *Occupational Medicine (London)*, Vol.53, PP. 95-101.

Galy, E.; Cariou, M. & Mélan, C. (2012). What is the relationship between mental workload factors and cognitive load types?, *International Journal of Psychophysiology*, Vol.83, No.3, pp. 269-275

Galy, E.; Mélan, C. & Cariou, M. (2008). Investigation of task performance variations according to task requirements and alertness across the 24-h day in shift-workers. Ergonomics, Vol.51, No.9, pp. 1338-51

Galy, E.; Mélan, C. & Cariou, M. (2010). Investigation of ATCs' response strategies in a free recall task: what makes auditory recall superior to visual recall? *International Journal of Aviation Psychology*, Vol.20, No.3, pp. 295-307.

Glenberg, A. M., & Swanson, N. G. (1986). A temporal distinctiveness theory of recency and modality effects. *Journal of Experimental Psychology: Learning, Memory, and Cognition*, Vol.12, No1, pp. 3-15

Gronlund, S.D.; Dougherty, M.R.P., Durso, F.T., Canning, J.M. & Mills, S. H. (2005). Planning in air traffic control: Impact of problem type. *International Journal of Aviation Psychology*, Vol.15, pp. 269-293

Gronlund, S.D.; Ohrt, D.D., Dougherty, M.R.P., Perry, J.L. & Manning, C.A. (1998). Role of memory in air traffic control. *Journal of Experimental Psychology: Learning, Memory, & Cognition*, Vol.15, pp. 846-858

Harvey, A.J. & Beaman, C.P. Input and output modality effects in immediate serial recall. *Memory*, Vol.15, No.7, pp. 693-700

Kecklund, G., Akerstedt, T. & Lowden, A. (1997). Morning work: effects of early rising on sleep and alertness. *Sleep*, Vol. 20, pp. 215-223

Kalyuga, S., Ayres, P., Chandler, P. & Sweller, J. (2003). The expertise reversal effect. *Educational Psychologist*, Vol.38, pp. 23–31

Khaleque, A. (1984). Circadian rhythms in heart rate of shift and day workers. *Journal of Human Ergology*, Vol.13, pp. 23-29

Lille, F. & Cheliout, F. (1982). Variations in diurnal and nocturnal waking state in air traffic controllers. *European Journal of Applied Physiology*, Vol.49, pp. 319-328

Lorenzetti, R. & Natale, V. (1996). Time of day and processing strategies in narrative comprehension. *British Journal of Psychology*, Vol.87, pp. 209-221

Luna, T.D., French. J. & Mitcha, J.L. (1997). A study of USAF air traffic controller shiftwork: sleep, fatigue, activity, and mood analyses. *Aviation Space and Environmental Medicine*, Vol.68, pp. 18-23

Madigan, S. A. (1971). Modality and recall order interactions in short-term memory for serial order. *Journal of Experimental Psychology*, Vol.87, pp. 294-296

Means, B.; Mumaw, R.J., Roth, C., Schlager, M.S., McWilliams, E., Gagne, E. et al. (1988). *ATC training analysis study: Design of the next-generation ATC training system.* report N°. FAA/OPM 342-036, Washington, DC: Department of Transportation/federal Aviation Administration, USA

Mélan, C.; Galy, E. & Cariou, M. (2007). Mnemonic Processing in Air Traffic Controllers (ATCs): Effects of Task Parameters and Work Organization. *International Journal of Aviation Psychology*, Vol.17, No.4, pp. 391-409

Monk, T.H.; Fookson, J.E., Moline, M.L. & Pollak, C.P. (1985). Diurnal variation in mood and performance in a time isolated environment. *Chronobiology International*, Vol.2, pp. 185-193

Oakhill, J. & Davies, A.M. (1989). The effects of time of day and subjects' test expectations on recall and recognition of prose materials. *Acta Psychologica*, Vol.72, pp. 145-157

Oberauer, K. & Vockenberg, K. (2009). Updating of working memory: Lingering bindings. *The Quarterly Journal of Experimental Psychology*, Vol.62, pp. 967-987

Owens, D. S.; Macdonald, I., Tucker, P., Sytnik, N., Totterdell, P., Minors, D., et al. (2000). Diurnal variations in the mood and performance of highly practiced young women living under strictly controlled conditions. *British Journal of Psychology*, Vol.91, pp. 41-60

Penney, C. G. (1989). Modality effects and structure of short-term verbal memory. *Memory and Cognition*, Vol.17, No.4, pp. 398-422

Prizmic Z.; Vidadek S., Radosevic-Vidadek B. & Kaliterna L. (1995). Shiftwork tolerance and 24-h variations in moods. *Work Stress*, Vol. 9, pp. 327-334

Rose, R.M.; Jenkins, C.D., Hurst, M., Herd, J.A. & Hall RP. (1982). Endocrine activity in air traffic controllers at work. II. Biological, psychological and work correlates. *Psychoneuroendocrinology*, Vol.7, pp. 113-123

Rosa R. (1995). Extended workshifts and excessive fatigue. Journal of Sleep Research, Vol.4, pp. 51-56

Rummer, R. & Schweppe, J. (2005). Evidence for a modality effect in sentence retention. *Psychonomic Bulletin & Review*, Vol.12, No.6, pp.1094-1099

Prince, S.E., Daselaar, S.M. & Cabeza, R. (2005). Neural of relational memory: successful encoding and retrieval of semantic and perceptual associations. *Journal of Neuroscience*, Vol.25, pp. 1203-1210

Schneider, V.I.; Healy, A.F. & Barshi, I. (2004). Effects of instruction modality and read-back on accuracy in flowing navigation commands. *Journal of Experimental Psychology: Applied*, Vol.10, pp. 245-257

Schnotz, W. & Kürschner, C. (2007). A reconsideration of cognitive load theory. *Educational Psychology Review*, Vol.19, pp. 469–508

Siegrist, J. (2010). Effort-reward imbalance at work and cardiovascular diseases. *International Journal of Occupational Medicine and Environmental Health*, Vol.23, pp. 279-285

Sweller, J. (1988). Cognitive load during problem solving: effects on learning. *Cognitive Science*, Vol.12, pp. 257–285

Sweller, J. (1994). Cognitive load theory, learning difficulty and instructional design. *Learning and Instruction*, Vol.4, pp. 295–312

Sweller, J. & Chandler, P. (1994). Why some material is difficult to learn. *Cognition and Instruction*, Vol.12, pp. 185–233

Thayer, R.E. (1989). *The biopsychology of mood and arousal*. New-York: Oxford University Press

Tucker, P., Lombardi, D., Smith L. & Folkard, S. (2006). The Impact of Rest Breaks on Temporal Trends in Injury Risk. *Chronobiology International*, Vol. 23, pp. 1423–1434

Tucker P, Smith L, Macdonald I, Folkard S. (1998). Shift length as a determinant of retrospective on-shift alertness. *Scandinavian Journal of Work and Environmental Health*, Vol.24, pp. 49-54

Wright, K.P., Hull, J.T. & Czeisler, C.A. (2002). Relationship between alertness, performance, and body temperature in humans. *American Journal of Physiology*, Vol. 253, pp. 1370-1377

Wickens, C.D. & Hollands, J.G. (2000). *Engineering psychology and human performance* (3rd ed.). Upper Saddle River, NJ: Prentice Hall, USA

Chapter 5

Simulation of Team Cooperation Processes in En-Route Air Traffic Control

Kazuo Furuta, Kouhei Ohno, Taro Kanno and Satoru Inoue

Additional information is available at the end of the chapter 0

1. Introduction

Recent increase in air traffic demands makes the role of Air Traffic Control (ATC), which supports safety and efficiency of aviation, more important than ever. As aviation technologies have progressed, automation and computer supports are being introduced in cockpits, but ATC still heavily relies on human expertise of Air Traffic Control Officers (ATCOs). It is therefore necessary to understand ATC tasks from a viewpoint of ATCOs' cognitive behaviour in order to assess and improve task schemes and training programs for ATC.

There are a few different phases in ATC, but this study exclusively focuses on en-route ATC. En-route ATC is performed by a team of ATCOs usually made up with a Radar Controller (RC) and a Coordination Controller (CC). RC monitors the radar screen, communicates with air pilots by radio, and gives instructions to them, while CC makes coordination with CCs in charge of other sectors, and supports RC. Team cooperation is therefore a key issue for good control performance, and study on team cooperation processes is an important issue.

We have been studying team cooperation processes in en-route ATC based on ethnographic field observation, and already proposed a cognitive model of team cooperation in en-route ATC as shown in Fig. 1 (Furuta et al., 2009; Soraji et al., 2010). In this model, establishment of Team Situation Awareness (TSA) on the air traffic is a key process for smooth cooperation. TSA here can be defined as a combination of individual situation awareness (Endsley, 1995) and mutual beliefs on it (Shu & Furuta, 2005). Once TSA has been established, individual tasks will be planned and executed almost implicitly, and TSA almost determines decision by ATCOs. The cognitive processes of a controller team after TSA acquisition are well described by a distributed version of the recognition-primed decision model (Klein, 1997).

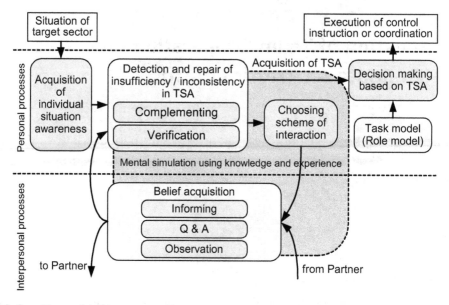

Figure 1. Cognitive model of team cooperative processes

Though the proposed cognitive model can explain the primary features of team cooperation in en-route ATC, it is still static and cannot explain dynamics of the processes. The aim of this research is to study detailed cognitive processes and communication strategies for establishing TSA using computer simulation. Computer simulation has been a useful tool in cognitive systems engineering for validation and sophistication of cognitive models (e.g., Furuta & Kondo, 1993; Cacciabue & Cojazzi, 1995; Cacciabue, 1998; Chang & Mosleh, 2007; NASA, 2011), because no ambiguities are allowed in coding executable computer programs for simulation. Interaction schemes and communication strategies for establishing TSA were discussed in our previous study, and computer simulation is a good approach also to reveal how ATCOs organize these schemes and strategies in their actual field settings.

2. Theoretical backgrounds

This study as well as our previous study relies on the Mutual Belief Model (MBM) as the theoretical basis. MBM is represented as a three-layered structure of items believed by team members, and it is a framework to explain how a team of individuals can coordinate their tasks smoothly to achieve shared team goals (Kanno, 2007). It is premised in MBM that there is a layered structure of human beliefs for all cognitive constructs and that establishing its consistency enables one to cooperate his/her partners. Figure 2 illustrates MBM for a team with two members, A and B. The layered structure theoretically repeats ad infinitum, but considering the first three layers will be enough in the real situation.

The first layer is the place to describe ones own cognition: what are perceived, recognized, believed, predicted, intended, planed, and so on. The second layer is for describing ones beliefs on his/her partner's cognition, and it reflects partner's first layer. The third layer describes ones beliefs on partner's beliefs on his/her own cognition, which are the self-image through the partner. Information not only on the environment but also on partner's cognition is used in team cooperation, and all of them are described in the layers of beliefs in MBM.

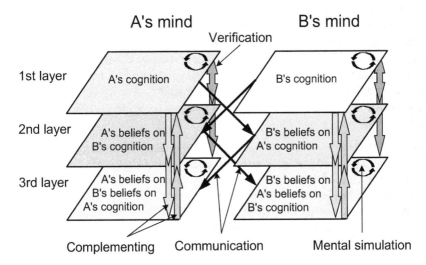

Figure 2. Mutual Belief Model and interaction schemes

Each belief item in this structure is obtained through internal or interpersonal interactions as well as through perception directly from the environment. There are four types of interactions: verbal and nonverbal communication, mental simulation (inference), complementing (presumption), and verification (consistency check). Communication is a process to transfer some belief item from one person to another by explicit utterance or observation. Mental simulation is a process to derive new belief items from some others within the same MBM layer by inference. In complementing, some belief item will be copied from one MBM layer to another within the same person. Verification is comparison of belief items between different MBM layers within the same person to check consistency among mutual beliefs.

Manifestation styles of interpersonal interactions are threefold: informing, query and answer (Q&A), and observation. Informing is an interaction to transfer some information from one person to another. This interaction is carried out by writing into a flight data strip as well as by verbal communication in ATC. Q&A is a combination of requesting information and corresponding reply to the request. This interaction is carried out almost verbally and it is used primarily for verification. Query is also used for the purpose of informing, and no replies are returned often in such a case. Observation is guessing partner's cognitive processes or mental states from his/her performance. The observed obtains a belief that he/she is being observed in most cases.

3. Review of field observation

3.1. Field observation

This work is based on the same data from the field observation of our previous work [1, 2]. The field observation was performed at the Tokyo Area Control Centre from 7 to 11 of May, 2007 during time periods of around three hours a day with relatively heavy traffic in the daytime so that the traffic imposed a certain level of workload on ATCOs. Different RC-CC pairs who were on a shift for the target sector were observed, but neither other sectors nor shift supervisors were observed. The target of observation was a sector called "Kanto-north" (T03), which spreads over the northern area of Tokyo. A lot of air traffic that departs from and arrives at two hub airports, the Tokyo International Airport (HND) and the Narita International Airport (NRT), smaller airports, and Air Force Bases (AFBs) passes through this sector.

Behavior of ATCOs was vide-recorded with two home video cameras, and another one recorded the radar screen of an auxiliary console where the same screen image was displayed that RCs were monitoring. An IC recorder attached on the controller console above the radar screen was used to record conversation between ATCOs. Flight data and radio communication records were also provided from the control centre. Combined video-audio records were made from the audio data and the video data of the radar screen synchronizing their time stamps. Radio communication and conversation between ATCOs were then transcribed, and speakers and listeners of conversation were identified. Actions of ATCOs were next recognized from the video data and added to the transcribed protocol data.

With the help of a rated ATCO, we segmented the protocol data by the basic unit of ATC instruction, clarified relations between segments, and identified expert knowledge and judgment behind them. Distributed cognitive processes among ATCOs were then inferred and reconstructed from the field data following the classification framework of MBM interactions. First we did not classify utterances themselves into particular categories but focused on reconstruction of MBM interactions. When we constructed a team cooperation model afterwards, we considered correspondence between the both and interaction schemes to classify utterances considering manifestation of communication.

3.2. Reclassification of interactions

The classification of interactions given in the previous chapter is a little ambiguous to describe precise processes of team cooperation; we will modify the classification slightly. Since interactions are basically information transfer from somewhere to somewhere, they are classifiable by the origin and the destination of information transfer. Table 1 summarizes the new classification.

Perception is a special interaction that the perceiver acquires information actively or passively from his/her working environment rather than another person. The acquired information is added into the first layer of the perceiver. Transmission (informing) is a verbal communication to transfer information from the speaker to the listener. If the speaker talks his/her own cognition, the information is transferred from speaker's first layer to listener's second layer. If the speaker talks his/her belief on listener's cognition, the information is transferred from speaker's second layer to listener's third layer. Observation is a nonverbal communication of observing partner's behavior to get belief on his/her belief in observer's second layer. Completion is a subcategory of complementing in the previous chapter. It is a process to accept what is believed by ones partner as his/her own belief; it is achieved by copying the belief item from the second to the first layer. Assumption is another subcategory of complementing, where one assumes that what he/she said to the partner is accepted. It is achieved by copying the belief item from the third to the second layer. Completion and assumption are mirroring processes between the speaker and the listener. Inference is the same process as the mental simulation in the previous chapter that derives new belief items from some others within the same layer. Query is a request of transmission for verification of the consistency of belief items in different layers.

Interaction	Type	Origin	Destination
Perception	Personal	Environment	1st layer
Transmission	Interpersonal	1st layer	2nd layer
		2nd layer	3rd layer
Observation	Interpersonal	1st layer	2nd layer
Completion	Personal	2nd layer	1st layer
Assumption	Personal	3rd layer	2nd layer
Inference	Personal	1st layer	1st layer
		2nd layer	2nd layer
		3rd layer	3rd layer

Table 1. Classification of interactions

3.3. Visualizing analysis results of field data

In order to exactly describe interactions between ATCOs including internal cognitive processes, we defined several classes of mental constructs, which are basic and unit description of beliefs in ATCO's MBM structure. Using the mental constructs, the results of cognitive task analysis of our previous work were transcribed down in a formal expression similar to predicate. The following are primary mental constructs and their meanings.

focus(Aircraft1, Aircraft2, ...)
   Attend and handle Aircraft1, Arcraft2, and so on in a group.
priority(Aircraft1, Aircraft2, ...)
   The priority among Aircraft1, Aircraft2, and so on is in this order.
instruction(Aircraft, Parameter, Value)
   Give Aircraft an instruction to keep Parameter at/below/above Value.
constraint(Aircraft, Parameter, Value)
   Consider a constraint for Aircraft to keep Parameter at/below/above Value.
execute(Action, Arg1, Arg2, ...)
   Execute an Action with arguments of Arg1, Arg2, and so on.

Describing analysis results in such a representation enables one to compare them easily with those of computer simulation.

3.4. Features of team interactions between ATCOs

In this study, we first developed a tool for visualizing the analysis results of the previous work on team interactions for detailed review. The same tool is also used later to visualize simulation results. The tool reads a file of formal descriptions of analysis or simulation results, and chronologically visualizes interactions between ATCOs by animating moves of information between MBM layers. Consequently, four features of team interactions have been found from the visualization. These features provide valuable hints and justifications in constructing a simulation model of team cooperation processes.

The first feature is the dominance of RC, which means that RC's cognition often starts a sequence of team cognitive processes. Perception by RC and succeeding transmission of the perceived information are frequently observed in the field data. Observation by CC on RC's behaviour is substitutive for transmission by RC in the above. Information transfer from RC to CC is more frequent than that in the opposite direction. Questioning by RC to CC is sometimes observed, but it is primarily for informing CC of RC's thought to request CC's support rather than literally asking CC's thought. Team cognitive processes are thereby paced by RC's cognition.

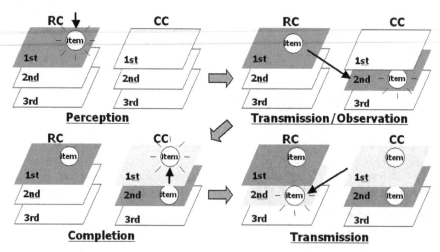

Figure 3. Interaction pattern for efficient TSA development

Secondly a recent topic is likely to be focused on in team interactions and deliberated further. It is the so-called recency effect, which is frequently observed in general cognitive processes.

The third finding is that an efficient pattern of TSA development often appears. Figure 3 shows a case where the pattern starts from perception of something by RC. The perceived item is transferred to the second layer of CC by transmission or observation in the next moment. CC then copies the item to the first layer by completion and informs RC of his/her belief to make another copy of the item in RC's second layer. Having finished all these steps successfully, the team has copies on the same belief item in both the first and the second layer, and they form sound TSA. The complementary pattern exists that starts from perception by CC.

Finally, after having acquired TSA, ATCOs will start interactions for deepening their thoughts based on the acquired TSA like considering strategies for the focused issue.

4. Simulation model

4.1. Flow of simulation

Computer simulation of interactions between RC and CC has been developed. Each member of the team is modelled as an agent in the simulation system: RC agent and CC agent. The both agents repeat choosing and executing a cognitive task to modify beliefs by referring to its own mutual belief module. Figure 4 illustrates the flow of simulation.

Each agent first looks at the mutual belief module and create a list of cognitive tasks applicable to the current situation. A cognitive task is represented as a production rule, which consists of the condition part and the action part. The general rules related to basic interactions for establishing TSA are hard-coded in the agents, and their implementation is explained below.

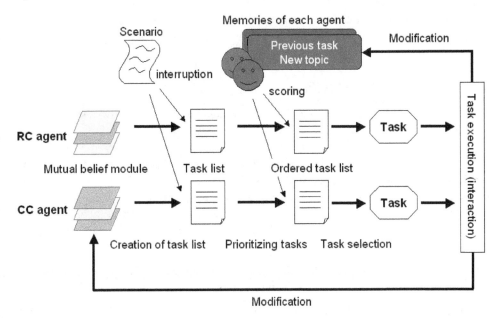

Figure 4. Flow of simulation

Observation (obs): If an agent executes any task visible to the partner agent, the latter will see the task and create the corresponding belief item in its second layer.

Completion (com): If any belief item contained in the second layer of an agent is missing in the first layer, the agent tries to copy the item into the first layer.

Assumption (ass): If any belief item contained in the third layer of an agent is missing in the second layer, the agent tries to copy the item into the second layer.

Inference (inf): A production rule fetched from the rule base is applied, if the rule is applicable to the present situation. Since this simulation exclusively focuses on team cooperation processes, it does not deal with the domain expertise specific to ATC explicitly. Expert judgment and inference are modelled as simple application of predefined rules that are very specific to each simulation case.

Perception (per): All perception tasks are predefined in the simulation scenario and they are triggered by time as interruption. Simulation scenarios are to be set up based on the field observation data, if simulation is to be done for the situations observed in the field.

Transmission (trn): If any belief item in agent's first layer does not exist in its second layer, the agent attempts to inform the item to the partner agent, i.e., tries to copy the item into the second layer of the partner agent. In execution of transmission, the transmitter agent also copies the item also into the third layer of its own, assuming that transmission is successfully done.

There is another kind of cognitive task, execution (exe), which is to execute some control action like hand-off, point-out, or giving a control instruction to a pilot. Execution is triggered as a result of rule application by inference. Query (qry) is not yet implemented in the present simulation model.

4.2. Prioritization of cognitive tasks

The agents next prioritize cognitive tasks in the created task lists. Cognitive tasks are scored referring not only to the basic score predefined for each task type but also to the past records of simulation hold in the memory models. When a cognitive task is triggered, its basic score shown in Table 2 is first given. The recency effect observed in the field data can be taken into consideration by adding a bonus if the task is related to the belief item created in the previous simulation step. If a particular sequence appeared between the successive tasks, another bonus will be added, which promotes the interaction pattern for efficient TSA development discussed in the previous chapter. After scoring, the task with the highest score is chosen and executed.

The scores have been adjusted so that simulation can well replicate the interactions observed in the actual ATCOs, and Table 2 lists the final scores.

The degree of match between the analysis of observation data and the simulation results was evaluated in three levels. If simulation could predict exactly a task appeared in the field data, it is labelled "perfect match." An internal task that is unobservable from the outside of an individual but predictable will be labelled "predictive match," if it is included in the simulation results. "Essential match" means that a pair of tasks in the observation data and the simulation results has the same effects on mutual beliefs, though they are different in appearance.

Task type or task sequence	Score	Explanation
Perception (per)	600	Basic score
Transmission (trn)	100	Basic score
Observation (obs)	0	Basic score
Completion (com)	100	Basic score
Assumption (ass)	100	Basic score
Inference (inf)	200	Basic score
Execution (exe)	500	Basic score
Transmission → Assumption	300	Bonus for efficient TSA development
Perception → Transmission	100	Bonus for efficient TSA development
Inference → Transmission	100	Bonus for efficient TSA development
Transmission → Completion	100	Bonus for efficient TSA development
—	100	Bonus for the recency effect

Table 2. Scores for prioritizing cognitive tasks

A measure of TSA appropriateness is completeness, which is defined as the ratio of belief items shared by the two corresponding belief layers of a team over the total number of belief items in the mirrored layer of the pair [4]. In these case studies, completeness of the second MBM layers was evaluated, and one should notice that just belief items relevant to related aircrafts were considered in simulation. Consequently, the results are more extreme than in the real situation, where ATCOs conceive belief items irrelevant to related aircrafts. Another measure, soundness, was ignored in this study, because no error mechanism was considered in the simulation model of this work and the soundness of TSA will be almost 100% in any case.

5.1. Case 1 (P.M, may 7, 2007)

In the first case (Fig. 5), the controller team dealt with a situation where three aircrafts bound for NRT from the northwest, the first one of which is denoted ac1 below, and another one from the north, ac2, was entering the sector almost at the same time. In addition, a warplane, ac3, were approaching from the southwest to cross over the flight path towards NRT, and it might interfere one of the four aircrafts. The controller team discussed and decided the order of aircrafts descending toward NRT while managing separation of the warplane from the others.

Table 3 compares simulated and observed interactions in this case. Items printed in italics were given in the simulation scenario and those in bolds show matched items between simulation and observation.

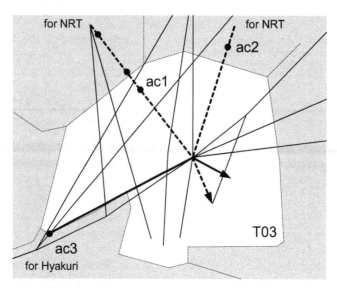

Figure 5. Situation of traffic in Case 1

Having watched the radar screen, RC judged the priority between ac1 and ac2 as ac1 should go first. CC recognized that RC was considering the order of ac1 and ac2 through observation of RC, CC asked a question on RC's judgment in the field, while RC transmitted its judgment explicitly to CC in simulation. RC then recognized that ac3 would interfere the other aircrafts, and CC proposed to instruct ac3 to go down in advance. RC agreed and proposed a particular altitude of descent as FL130, but CC came to know that ac3 should be kept above FL170 as a result of coordination with Hyakuri AFB control. Meanwhile, RC continued monitoring the positions of aircrafts, and CC observed these actions. CC proposed to change the order of aircrafts as ac2 should go first, and transmitted this judgment to RC.

Among 26 interactions that were extracted from the field data, four items were used to setup the simulation scenario, and simulation successfully replicated 18 items: 81.8% (18/22) of interactions observed in the field data. The order of appearance of interactions was almost the same. Two queries observed in the field were simulated as transmissions, and two transmissions by CC were not simulated. Four transmissions by RC predicted by simulation were not executed actually, and no assumptions were observable in the field.

The assessment result of TSA is shown in Fig. 6. Completeness of both RC and CC sometimes degrades, when one of them perceived new information or generated new belief items by inference. It was however recovered soon by verbal communication. TSA completeness of CC was slightly better than that of RC, but difference was very small, because they kept close verbal communication for deciding the priority of related aircrafts.

Simulation		Observation	
Step	Actor and action	Time	Actor and action
0	*RC: per - focus(ac1, ac2)*	14:07:40	*RC: per - focus(ac1, ac2)*
	CC: obs - focus(ac1, ac2)*		**CC: obs - focus(ac1, ac2)***
1	RC: trn - focus(ac1,ac2)		**RC: inf - priority(ac1, ac2)+**
	CC: com - focus(ac1, ac2)+	14:07:47	CC: qry - priority(ac1, ac2)^
2	RC: ass - focus(ac1, ac2)		**CC: com - focus(ac1, ac2)+**
3	**RC: inf - priority(ac1, ac2)+**		**CC: com - priority(ac1, ac2)+**
4	**RC: trn - priority(ac1, ac2)^**	14:08:09	CC: trn - priority(ac1, ac2)
5	RC: ass - priority(ac1, ac2)		
	CC: com - priority(ac1, ac2)+		
6	*RC: per - focus(ac1, ac2, ac3)*		*RC: per - focus(ac1, ac2, ac3)*
	CC: obs - focus(ac1, ac2, ac3)*		**CC: obs - focus(ac1, ac2, ac3)***
7	RC: trn - focus(ac1, ac2, ac3)		
	CC: com - focus(ac1, ac2, ac3)+		**CC: com - focus(ac1, ac2, ac3)+**
8	RC: ass - focus(ac1, ac2, ac3)		
	CC: inf - instruct(ac3, alt down)+	14:08:22	**CC: inf - instruct(ac3, alt, down)+**
9	**CC: trn - instruct(ac3, alt, down)***		**CC: trn - instruct(ac3, alt, down)***
10	**RC: com - instruct(ac3, alt, down)+**		**RC: com - instruct(ac3, alt, down)+**
	CC: ass - instruct(ac3, alt, down)		
11	**RC: inf - instruct(ac3, alt, 130)+**		**RC: inf - instruct(ac3, alt, 130)+**
12	**RC: trn - instruct(ac3, alt, 130)***	14:08:48	**RC: trn - instruct(ac3, alt, 130)***
13	AC: ass - instruct(ac3, alt, 130)		**CC: com - instruct(ac3, alt, 130)+**
	CC: com - instruct(ac3, alt, 130)+		CC: trn - instruct(ac3, alt, 130)
17	*CC: per - constraint(ac3, alt, 170)*	14:08:57	*CC: per - constraint(ac3, alt, 170)*
18	**CC: trn - constraint(ac3, alt, 170)***	14:09:51	**CC: trn - constraint(ac3, alt, 170)***
19	*RC: per - focus(ac1', ac2')*		*RC: per - focus(ac1', ac2')*
	CC: ass - constraint(ac3, alt, 170)		
20	**RC: trn - focus(ac1', ac2')^**	14:10:01	**RC: qry - focus(ac1', ac2')^**
21	AC: ass - focus(ac1', ac2')		
	CC: com - focus(ac1', ac2')+		**CC: com - focus(ac1', ac2')+**
22	**RC: com -constraint(ac3, alt, 170)+**		
	CC: inf - priority(ac2, ac1)+	14:10:27	**CC: inf - priority(ac2, ac1)+**
23	**CC: trn - priority(ac2, ac1)***	14:10:55	**CC: trn - priority(ac2, ac1)***
24	**RC: com - priority(ac2, ac1)+**		**RC: com - priority(ac2, ac1)+**
	CC: ass - priority(ac2, ac1)		**RC: com - constraint(ac3, alt, 170)+**

Italics: Items given in the simulation scenario.
Bolds: Items matched between simulation and observation.
(* Perfect match + Predictive match ^ Essential match)

Table 3. Comparison of simulated and observed interactions in Case 1

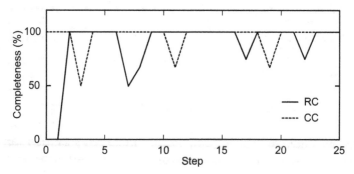

Figure 6. Completeness of the second MBM layers in Case 1

5.2. Case 2 (P.M. may 11, 2007)

In the second case (Fig. 7), an aircraft, ac1, departed from Yokota AFB and flied along the southernmost boundary of Sector T03 eastward to the Pacific Ocean. Many aircrafts departed from HND and NRT passed through the area during this time period; ATCOs had to concern about traffic interference. Another aircraft, ac2, departed from HND was climbing northward. Since the approach areas of NRT, HND and Yokota as well as neighboring sectors overlap in this area, demands for coordination with other sectors are relatively high.

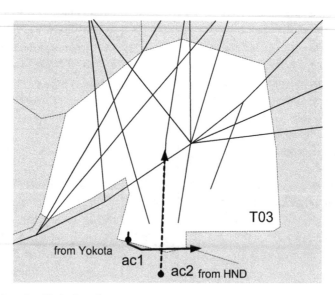

Figure 7. Situation of traffic in Case 2

Table 4 compares simulated and observed interactions in Case 2. At first, RC recognized the necessity of pointing out ac1 to HND control. CC who feared interference between ac1 and ac2 intended to lower ac2 to fly below ac1. Since ac2 still was under the control of HND, CC made coordination with HND departure control to limit its altitude below FL130. Ac1 would enter the control area of NRT due to its low altitude and CC recognized the necessity of pointing out ac1 to NRT control. CC communicated this possibility to RC. Having heard this communication, RC verified CC by a query with which control of NRT to make coordination: center (C) or departure (D) control. This query in the field data was substituted with transmission and completion in simulation.

	Simulation		Observation	
Step	Actor and action	Time		Actor and action
0	*RC: per - point_out(ac1, HND)*	12:08:40		*RC: per - point_out(ac1, HND)*
1	**RC: trn - point_out(ac1, HND)***			**RC: trn - point_out(ac1, HND)***
2	RC: ass - point_out(ac1, HND)			
	CC: com - point_out(ac1, HND)+			**CC: com - point_out(ac1, HND)+**
3	*CC: per - focus(ac1, ac2)*			*CC: per - focus(ac1, ac2)*
4	**CC: inf - coordinate(ac2, alt)+**			**CC: inf - coordinate(ac2, alt)+**
5	**CC: trn - coordinate(ac2, alt)***	12:09:10		**CC: trn - coordinate(ac2, alt)***
6	**RC: com - coordinate(ac2, alt)+**	12:09:26		**CC: trn - focus(ac1, ac2)***
	CC: ass - coordinate(ac2, alt)			**RC: com - focus(ac1, ac2)+**
7	**CC: exe - coordinate(ac2, alt)***			RC: trn - focus(ac1, ac2)
8	**CC: trn - focus(ac1, ac2)***			**RC: com - coordinate(ac2, alt)+**
9	**RC: com - focus(ac1, ac2)+**	12:09:31		**CC: exe - coordinate(ac2, alt)***
	CC: ass - focus(ac1, ac2)			CC: trn - coordinate(ac2, alt, 130)
10	*RC: per - enter(ac1, NRT)*			RC: com - coordinate(ac2, alt, 130)
11	**RC: trn - enter(ac1, NRT)***			*RC: per - enter(ac1, NRT)*
12	RC: ass - enter(ac1, NRT)	12:12:24		**RC: trn - enter(ac1, NRT)***
	CC: com - enter(ac1, NRT)+			**CC: com - enter(ac1, NRT)+**
13	**CC: inf - point_out(ac1, NRT)+**			**CC: inf - point_out(ac1, NRT)+**
14	**CC: trn - point_out(ac1, NRT)***			**CC: trn - point_out(ac1, NRT)***
15	**RC: com - point_out(ac1, NRT)+**			**RC: com - point_out(ac1, NRT)+**
	CC: ass - point_out(ac1, NRT)			
16	**RC: inf - point_out(ac1, NRT, C)+**			**RC: inf - point_out(ac1, NRT, C)+**
17	**RC: trn - point_out(ac1, NRT, C)^**			**RC: qry - point_out(ac1, NRT, C)^**
18	RC: ass - point_out(ac1, NRT, C)			
	CC: com - point_out(ac1, NRT, C)			
19	**CC: inf - point_out(ac1, NRT, D)+**			**CC: inf - point_out(ac1, NRT, D)+**
20	**CC: trn - point_out(ac1, NRT, D)***	12:12:31		**CC: trn - point_out(ac1, NRT, D)***
21	**RC: com - point_out(ac1, NRT, D)+**			**RC: com - point_out(ac1, NRT, D)+**
	CC: ass - point_out(ac1, NRT, D)			RC: trn - point_out(ac1, NRT, D)

Italics: Items given in the simulation scenario.
Bolds: Items matched between simulation and observation.
(* Perfect match + Predictive match ^ Essential match)

Table 4. Comparison of simulated and observed interactions in Case 2

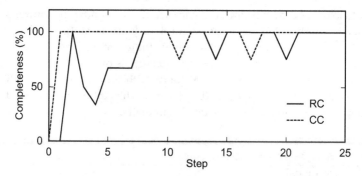

Figure 8. Completeness of the second MBM layers in Case 2

Among 25 interactions extracted from the field data, three items were predefined in the simulation scenario, and simulation could predict 17 items, 77.3% (17/22) of the observed. The query observed in the field was substituted with transmission and completion, and no assumptions were observable in the field. Three transmissions and one completion were not simulated.

Figure 8 shows completeness of TSA in Case 2. Dominance of RC appears here, that means, RC leads team cognitive processes and CC follows by eagerly obtaining mutual beliefs on them. Consequently, completeness of TSA is a little better for CC than RC.

5.3. Case 3 (P.M. may 11, 2007)

The third case shows how RC and CC resolved conflict on control strategy. An aircraft, ac1, which departed from Hyakuri AFB, might interfere with another aircraft, ac2, bound for HND (Fig. 9). CC made coordination with Hyakluri AFB control before hand-off for ac1 to fly north around a fix with identification code GOC. RC, however, instructed ac1 to fly south around GOC. Having heard RC's instruction, CC noticed RC's intention and recovered from the conflicting state of mutual beliefs.

Table 5 compares simulated and observed interactions in the last case. CC expected that RC would give ac1 an instruction to fly north around GOC and made coordination with Hyakuri AFB control for its preparation. RC, however, thought the southern rout is better and gave an instruction to fly south around GOC. CC recognized RC's intention by observation and remedied the conflicting mutual belief.

There were 13 interactions from the field data, two of them were used for the simulation scenario, and 9 of them appeared in the simulation result. The hit rate was 81.2% (9/11). Simulation missed one transmission and one inference, and no assumptions were observable in the field data.

Figure 10 shows the assessment result of TSA completeness. Completeness of TSA was worse for both RC and CC in this case than in the other cases, because there was a conflict on control strategy between RC and CC. Finally CC recognized the conflict by observing

RC's instruction to the pilot, and CC's completeness recovered at the end of simulation. RC, however, did never recognize that CC had conceived a different strategy, and RC's completeness remained low till the end of simulation. Completeness of TSA thereby can be a good measure that exactly shows whether or not the members of a cooperating team share situation awareness.

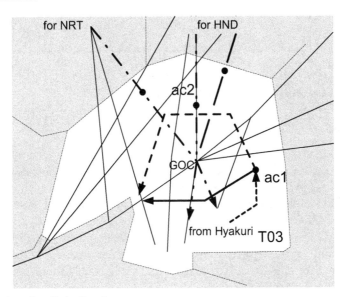

Figure 9. Situation of traffic in Case 3

	Simulation		Observation	
Step	Actor and action	Time		Actor and action
0	*CC: per - focus(ac1, ac2)*			*CC: per - focus(ac1, ac2)*
1	**CC: inf - instruction(ac1, dir, N)+**	14:05:27		**CC: inf - instruction(ac1, dir, N)+**
2	**CC: trn - focus(ac1, ac2)***	14:06:26		**CC: trn - focus(ac1, ac2)***
3	**RC: com - focus(ac1, ac2)+**			**RC: com - focus(ac1, ac2)+**
	CC: ass - focus(ac1, ac2)			**RC: inf - instruction(ac1, dir, S)+**
4	**RC: inf - instruction(ac1, dir S)+**	14:08:27		RC: trn - focus(ac1, ac2)
6	*CC: per - hand_off(hyakuri, ac1)*	14:09:17		*CC: per - hand_off(hyakuri, ac1)*
7	**CC: trn - hand_off(hyakuri, ac1)***	14:09:53		**CC: trn - hand_off(hyakuri, ac1)***
8	**RC: com - hand_off(hyakuri, ac1)+**			**RC: com - hand_off(hyakuri, ac1)+**
	CC: ass - hand_off(hyakuri, ac1)	14:10:22		**RC: exe - instruction(ac1, dir, S)***
9	**RC: exe - instruction(ac1, dir, S)***			**CC: obs - instruction(ac1, dir, S)***
	CC: obs - instruction(ac1, dir, S)*			**CC: com - instruction(ac1, dir, S)+**
10	**CC: com - instruction(ac1, dir, S)+**			CC: inf - instruction(ac1, dir, S)

Italics: Items given in the simulation scenario.
Bolds: Items matched between simulation and observation.
(* Perfect match + Predictive match ^ Essential match)

Table 5. Comparison of simulated and observed interactions in Case 3

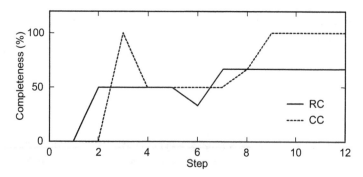

Figure 10. Completeness of the second MBM layers in Case 3

5.4. Discussion

Simulation could successfully replicate most of the interactions, around 80% in all cases, observed in the field. Table 6 shows a summary of match between simulated and observed interactions. This good match indicates that the simulation model of team interactions including detailed implementation was appropriate. The details of the model here include the initiation conditions and the prioritization scheme of cognitive tasks, which are required for constructing an executable simulation program. Assumptions, which are completely unobservable from the third party, were simulated but not observed in the field. Human expertise required for en-route ATC is beyond the scope of this simulation, because it is represented as simplistic production rules.

Completeness of CC's second layer outperformed that of RC's second layer in all cases. It means that CC is relatively eager to obtain mutual beliefs compared with RC by following and monitoring RC's actions. In contrast, RC is relatively independent from CC in deciding control actions and their timing. It resulted in dominance of RC observed in the field. Interactions like perception and inference generate new belief items in MBM layers to lower completeness of TSA, but it is usually recovered immediately by some sort of communication. These tasks that lower completeness of TSA, however, contribute to deepen thought on the current issues. It seems a standard style to repeat such a tandem process of deepening thought and establishing TSA in cooperation of ATCOs.

Number of interactions	Case 1	Case 2	Case 3
Simulation scenario	4	3	2
Perfect match	6	7	4
Predictive match	12	10	5
Essential match	2	1	0
Not simulated	2	4	2
Observed interactions	26	25	13

Table 6. The degree of match between observed and simulated interactions

6. Conclusion

The results of analysis on field observation data in en-route ATC were transcribed using predefined cognitive constructs and visualized to reveal typical patterns of team interactions for establishing TSA. Computer simulation then has been developed based on the cognitive model of team cooperation processes of our previous study as well as the revealed typical interaction patterns, and simulation was performed for three cases from the field observation. Simulation could replicate around 80% of ATCOs' interactions and the typical features of interactions observed in the field. The good match indicates that the cognitive model of team cooperation processes proposed in our previous study has a reality. In addition, simulation can explain the cognitive mechanisms of team cooperation processes for not only verbal but also for non-verbal interactions.

Appropriateness of TSA was evaluated using the simulation results evaluating completeness of the second MBM layers. Dominance of RC observed in the field resulted in higher TSA completeness for CC than for RC. It is also shown that TSA completeness degrades when RC or CC obtains new belief items by perception or inference but it is soon recovered by verbal or nonverbal communication. Assessment of TSA by computer simulation is thereby a useful mean to visualize the degree of team cooperation.

The simulation model of this work still has many limitations though. Firstly, expertise required for en-route ATC was represented as simplistic production rules that are very specific to the cases to be simulated, and this implementation lacks generality. Secondly, this work considers no models of errors, and the interactions generated by this model are normative. Smooth and efficient team cooperation is certainly a key for error-free and safe ATC performance. It is expected but inconclusive from this work, however, if it leads also to a high throughput of the sector, because the observation data were obtained just for time periods with relatively heavy traffic in the daytime and no critical situations happened during the observation. These issues have been left for future studies.

Author details

Kazuo Furuta, Kouhei Ohno and Taro Kanno
Department of Systems Innovation, The University of Tokyo, Japan

Satoru Inoue
Electronic Navigation Research Institute, Japan

7. References

Cacciabue, P.C. (1998). *Modeling and Simulation of Human Behaviour in System Control*, ISBN 3-540-76233-7, Springer, London, UK

Cacciabue, P.C. & Cojazzi, G. (1995). An integrated simulation approach for the analysis of pilot-aeroplane interaction, *Control Engineering Practice*, Vol.3, No.2, pp. 257-266, ISSN 0967-0661

Chang, Y.H.J. & Mosleh, A. (2007). Cognitive modelling and dynamic probabilistic simulation of operating crew response to complex system accidents, Part 1: Overview of the IDAC model, *Reliability Engineering and System Safety*, Vol. 92, No.8, pp. 997-1013, ISSN 0951-8320

Endsley, M.R. (1995). Towards Theory of Situation Awareness in Dynamic System, *Human Factors*, Vol.37, No.1, pp. 32-64, ISSN 0018-7208

Furuta, K. & Kondo, S. (1993). An approach to assessment of plant man-machine systems by computer simulation of an operator's cognitive behavior, *International Journal of Man-Machine Studies*, Vol.39, No.3, pp. 473-493

Furuta, K.; Soraji, Y.; Kanno, T.; Aoyama, H.; Inoue, S .; Karikawa, D. & Takahashi, M. (2009). Cognitive Model of Team Cooperation in En-Route Air Traffic Control, *Proceedings of ESREL 2009*, pp. 1811-1816, ISBN 13 978-0-415-48513-5, Prague, Czech, September 2009

Kanno, T. (2007). The notion of sharedness based on mutual belief, *Proceedings of 12th International Conference on Human-Computer Interaction*, pp 1347-1351, ISBN 978-3-540-73738-4, Beijing, China, July 2007

Klein, G. (1997). Recognition-primed decision model: looking back, looking forward, In: *Naturalistic decision making*, C.E. Zsambok & G. Klein, (Eds.), pp. 285-292, ISBN 10 080581874X, Lawrence Erlbaum, Mahwah, NJ

NASA (2011). *MIDAS: Man-Machine Integration Design and Analysis System*, 15.07.2011, Available from: http://humansystems. arc.nasa.gov /groups/ midas/index.html

Shu, Y. & Furuta, K. (2005). An inference method of team situation awareness based on mutual awareness, *International Journal of Cognition, Technology, and Work*, Vol.7, pp. 272-287, ISSN 1435-5558

Soraji, Y.; Furuta, K.; Kanno, T.; Aoyama, H.; Inoue, S.; Karikawa, D. & Takahashi, M. (2010). Cognitive model of team cooperation in en-route air traffic control, *International Journal of Cognition, Technology, and Work*, 07.03.2012, Available from: http://www.springerlink.com/content/g6122786tjq7w81t/fulltext.pdf

Probability of Potential Collision for Aircraft Encounters in High Density Airspaces

R. Arnaldo, F.J. Sáez, E. Garcia and Y. Portillo

Additional information is available at the end of the chapter

1. Introduction

Collision risk estimation in airspace and mathematical modeling of mid-air collisions have been carried out for over more than 40 years [1]. During this period there has been a development of mathematical models for processes leading to possible collisions of aircraft flying nearby in order to estimate the risk of collision.

B. L. Marks [2] developed the principles in which a collision risk model could be developed in the early 1960s. Marks' work was modified and enhanced by P. Reich [3] and that model, later called the Reich model, has been the basis for many of the important developments in this field.

The Reich model uses information related to the probabilistic distributions of aircraft's lateral and vertical position, traffic flows of the routes, aircraft's relative velocities and aircraft dimensions to generate estimation of collision risk. However, this model does not cover adequately situations where ground controllers monitor the air traffic through radar surveillance and provide tactical instructions to the aircraft crews. Furthermore, the problem of collision risk modeling in the analysis of "high traffic density" ATC scenarios is different to that of "procedural scenarios", which have been developed by Reich [4] and Brooker [5], amongst others. This is mainly due to the active role of Controllers in the first case. In this case positive control is used extensively to modify the planned aircraft route. This requires the inclusion in the model of "human factor response" behavior.

These "collision risk models" were initially applied in the 60s to determine safe separation standards between pairs of aircraft flying at the same altitude on parallel courses over the North Atlantic Ocean [6]. Since then, new models have been developed and continually refined and improved. They have been applied for different geographic regions (USA [7], European airspace [8]), for oceanic or radar [9] environments, and different flight regimes (for example, high-altitude cruise and landing on close [10,11,12] and ultra close spaced runways [13,14]), for specific flight phases [15] (focused for example on the separation between aircraft on final approach and landing, when flight risks are greater than during any other phase of flight), for different types of separation (vertical, longitudinal and lateral) and also for current and future operational concepts [16], such as free flight [17], airborne self separation [18],….

Most of these models, amongst them the formula proposed by Brooker [19] for mid-air collision risk, involve the aggregation of terms comprising different factors related to: initiating events which produce defective flight paths; the probability of safety defenses correcting these defective flight plans; and traffic and kinematic scalers. But, as he indicates: "it does no more than spell out the mechanisms by which collisions logically have to occur. The hard problem is how to populate the parameters in the formulation with sensible numbers".

Risk models have also been developed for the estimation of conflict probability (understood as the probability that the distance between a pair of aircraft becomes smaller than some specified minimum separation value). Paielli and Erzberger's [20,21] emphasis was on the development of algorithms to numerically evaluate approximations of conflict probabilities. Prandini et al. [22,23] emphasized the analysis of the problem and distinguished three sub-problems of evaluating conflict probability.

The main point of conflict probability is its clear relation to a well known safety criterion in civil aviation: the separation minimum, which puts a requirement on the air traffic management system; not to let aircraft come closer to each other than a certain minimum distance. In addition to minimum separation values, ICAO (International Civil Aviation Organization) has also defined limiting criteria for acceptable risk levels of fatal accident, and in particular, for the risk of mid air collision [24]. The allowed probability values for such events are of the order of one mid-air collision or physical crossing per $10^{\wedge 9}$ flight hour.

Furthermore , some effort has been also devoted to the problem of aircraft conflict detection. An excellent survey of the different conflict detection and resolution schemes has been carried out by Kuchar [25,26], where the conflict detection schemes are classified according to the modeling method used for projecting the aircraft position in the future.

According to Brooker [9], mid-air collisions derived from radar inaccuracies are very rare, so to estimate their frequency, it is necessary to model the factors that might lead to such events. But this extremely low value makes it difficult to obtain reliable empirical results from reasonably computational amount of data.

As collisions are very unlikely events most of the previous approaches to estimate collision risk were centered on simulations techniques applicable to rare event estimations such as Montecarlo simulations [27,28]. Nevertheless, simulations are not enough, as the components of the collision models have to be verifiable, i.e. match reality, and cautious. 'Verifiable' in the present context means that the model description can be demonstrated to match what happens in practice, and that most of the parameters in the model can be measured directly by analyzing air traffic patterns.

Some authors, like Dr. L. Burt [29], have formulated expressions that attempt to estimate Pa, distinguishing four different aircraft encounters geometries. The mathematical formulas are customized for these geometries so they are only applicable for circumstances that they have been developed for. They barely provide an estimate of the average conditional probability of collision Pa but they do not provide an individual value of Pa for each encounter. Therefore, this approach does not assess the severity of each individual potential encounter.

Other authors, such as Campos [30] have calculated the probability of coincidence for aircraft on arbitrary straight flight paths (either climbing, descending, or in level flight) with constant speed as an upper bound for the probability of collision. Although in this approach the time and distance of closest approach are used to calculate the position for maximum probability of coincidence. In reference [31] same authors illustrate the relationship between the aircraft RMS (Root Mean Square) position error and the minimum separation distance for achieving a certain Target Level of Safety (TLS) for low probability of collision.

Nevertheless, most of the researches on this field have just worked in the estimation of probabilities of conflict (before deliberate actions are taken to solve the conflicts) and how these probabilities depend on aircraft separation standards. Different current and future Air Traffic Management operational concepts have been studied under this perspective in an attempt to reduce aircraft separation standards [32,33] or with the aim of designing proper avoidance maneuvers in order to maintain the prescribed minimum separation standards among aircraft [34,35].

The previous considerations give an idea of the complexity of using stored aircraft tracks, within a given scenario and time frame, to infer safety level, collision risk probability and associated system weaknesses. In most high density airspace scenarios recorded tracks can be obtained for all aircraft flying in it, for example, from Radar Data Processing systems (RDP). In fact, this provides us with a robust data source, which could be used for safety analysis. This could include indirect information which is closely related to the "human factor response". Despaite its importance not much effort have been devoted to the development of risk and collision models based upon the analysis of the stored aircraft tracks.

Furthermore, it has to be considered that the distribution of aircraft position errors over their intended tracks is one of the most important factors in determining route safety, and consequently it has been broadly studied. Reference [36], for instance, presents a modeling technique to compute the probability density function of position errors as the aircraft proceed along the route taking into account not only the time dependence, but also all the factors influencing an aircraft's position errors, e.g., surveillance and navigation errors, surveillance fix rate, and air traffic control procedures.

Following the research line initiated on [31,37,38] by the mentioned previous work, the authors are developing a more detailed mathematical model for both components of probability of collision in a radar ATC (Air Traffic Control) environment.

2. Fundamentals behind probability of collision estimation

Jaroslav Krystul [39] defines the risk as the probability of a particular adverse event occurring during a stated period of time. Usually, this is an event occurring when the system reaches a particular critical state. These events with a very small probability of occurrence are called rare events. Applying this definition to an ATC scenario, it is accepted that risk is closely related to those situations in which two aircraft are on conflict course and would not only pass closer than the prescribed horizontal and vertical separation minima but which would, in fact, collide.

The work presented here was originally inspired by the principle stated in [2] by B. L. Marks: "… *the task of relating collision risk to a traffic configuration can be taken in two parts:*

parts:

1. Determining the frequency with which aircraft are exposed to risk by passing close together; and
2. Determining what chance of collision is inherent in the passing".

According to this idea, the probability of aircraft collision can be expressed as:

$$P(collision) = FeR * P(pot.coll/pot.conf) * P(coll/pot.coll). \tag{1}$$

where:

- **FeR**, Frequency of exposition to Risk, here is considered as the relative frequency that an aircraft would potentially violate the separation standards defined for the particular situation, here referred to as potential conflict. It is easily seen that this value increases with the traffic density.
- **P(pot.coll/pot.conf)** is the conditional probability of a potential collision (pot.coll) between two aircraft that have previously violated the separation standards (pot. conf). Its value depends on the encounter kinematics and uncertainties associated to predicted positions. It represents the intrinsic severity of the encounter and it is independent of the traffic density.
- **P(coll/pot.coll)** is the conditional probability of collision among potential collisions having failed all the safety barriers (ATC, TCAS) which are in place to mitigate the risk.

A time horizon is established within which all aircraft positions are projected to explore existence of "potential conflicts". In the following discussion 10 minutes look ahead time has been considered. Accordingly, the relative frequency of potential collisions among potential conflicts F(pot.coll/pot.conf) could be expressed as:

$$F(pot.coll / pot.conf) = \frac{Num. \ of \ pot. \ collisions}{Num. \ of \ pot. \ conflicts} \approx E[P_a]. \tag{2}$$

where Num.pot.collisions is the number of aircraft that are about to collide (and will do if all safety barriers fail).

An initial expectation for probability of potential collision among potential conflicts, **E(Pₐ)**, could be obtained as the relative frequency that two aircraft, on a conflict course, would not only pass closer than the prescribed horizontal and vertical separation minima, but would in fact collide. This expression provides an expected, or global, value and does not assess the severity of each individual potential encounter itself. This chapter proposes an approach to estimate the severity of the encounter using the conditional probability of a potential collision P_a for each particular aircraft encounter. This proposed approach aims at improving the previous works by:

- Providing an individual probability of collision for each individual encounter based on the: (1) geometry and kinematics of the encounter, (2) the minimum predicted lateral separation at the CPA, and (3) the minimum predicted vertical separation at the CPA.
- Taking into consideration the radar data errors and the segmentation errors.

2.1. Consideration of aircraft protection zones

As stated by Ennis [41], a protected zone represents a region around a given aircraft that no other aircraft should penetrate.

A simplification of the Bellantoni [42] approach for the definition of a collision surface can be made by modelling the aircraft as a cylinder of diameter λ_{xy} and height λ_z as indicated in figure 1.

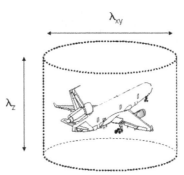

Figure 1. Aircraft representation

Two aircraft are taken as colliding if their cylinders touch. With this bounded and closed airspace region representing the aircraft, a "collision cylinder" can be defined as a larger cylinder of twice the dimensions represented in figure 1, and defined by height $2\lambda_z$ and radius $2\lambda_{xy}$ (see figure 2).

On the other hand, all high density traffic ATC scenarios have established minimum separation standards defined by two values, the minimum horizontal (R) and vertical (H) separations. When two aircraft are closer than these distances the ATC system is considered to have failed. These values (R, H) allow us to use another cylinder shaped protection model for all aircraft which should be free of any other aircraft to fulfil this separation minima (see figure 3). This volume will be called the "conflict cylinder" as it is considered that two aircraft potentially violating these separations are exposed to risk.

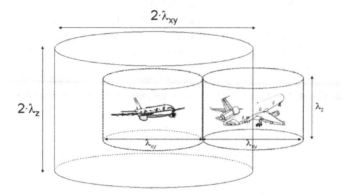

Figure 2. Collision cylinder definition

During the en route phase of flight, for example, the conflict cylinder would be 5 nm in radius and 2,000 ft in height. However, these current minimum separation standards were determined many years ago and the method by which they were calculated is not well documented. Recently, Reynolds & Hansman [42] identified factors involved in defining the aircraft separation standards and discussed the importance of accurate state information for controllers in maintaining them. Ennis & Zhao [43] examined the physical compositions of the protected zone and presented a formal approach to the analysis of minimum separation standards.

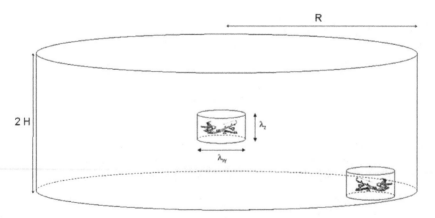

Figure 3. Conflict cylinder definition.

A summary of the modelling cylinders defined so far is presented in the following table.

When civil aircraft are climbing or descending, it is considered that pitch angles are small and so, vertical and horizontal dimensions have small changes. Therefore, all the "modelling cylinders" will be considered as horizontal, as indicated on figure 4.

As all the cylinders are considered parallel, the longitudes and surfaces ratios among them will be constant when they are projected onto any plane.

Cylinder	Diameter	Height
Aircraft representation	λ_{xy}	λ_z
Collision	$2\,\lambda_{xy}$	$2\,\lambda_z$
Conflict	2R	2H

Table 1. Modelling cylinders definition.

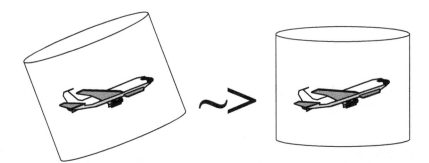

Figure 4. Modelling Cylinders Orientation

3. Derivation of a general expression for probability of collision (Pa)

In order to obtain a general expression of Pa an impact plane is defined as a generic projection plane containing the centre of reference aircraft ACi (assumed as static) and perpendicular to $\overrightarrow{v_{ji}}$ (relative velocity vector between the two aircraft i and j involved in the proximity event). Additionally, the collision area is defined as the projection, over the impact plane, of the collision cylinder ($2\,\lambda_{xy}$,$2\,\lambda_z$). If the conflict cylinder is settled in ACi, where its centroid is the one of the cylinder as well, the conflict area could also be defined as the projection of the conflict cylinder (2R, 2H). The CPAP (Closest Point Of Approach Projection) is a point with coordinates y1p and z1p obtained by projecting intruder aircraft. Figure 5 shows that a conflict will occur if ACj encounters the stationary conflict area, that is, if the CPAp coordinates (y1p, z1p) are inside the conflict area. In the same way, a collision will occur if ACj encounters the stationary collision area, that is, if the CPAp coordinates are inside the collision area.

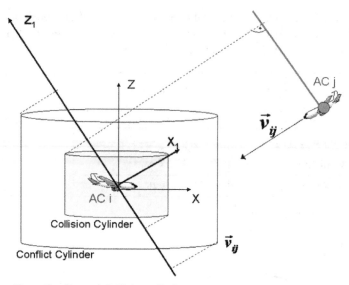

Figure 5. Impact Plane ,Conflict and Collision cylinder.

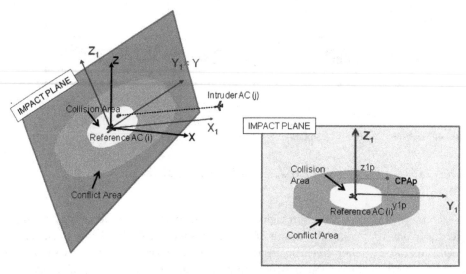

Figure 6. Impact Plane, Collision Area, Conflict Area and Projected CPA definition.

Considering the changes in the CPA coordinates due to radar and radar data segmentation errors, the probability of potential collision for an intruder aircraft that has violated the separation standards and whose projection consequently hits within the conflict area can be calculated as:

$$P_a(y_{1p}, z_{1p}) = \int_{S_{PCF}} dP_1 P_2 \approx S_{PCOL} \cdot \int_{S_{PCF}} f_1(y'_{1p} - y_{1p'}, z'_{1p} - z_{1p}) \cdot f_2(-y'_{1p} - z'_{1p}) dy'_1 dz'_1 \qquad (3)$$

This equation provides and individual probability of collision based on:

- geometry and kinematics of the encounter (S$_{PCOL}$),
- the predicted minimum lateral separation at the CPA (y$_{1p}$), and
- the predicted minimum vertical separation at the CPA (z$_{1p}$).

This takes into consideration the two probability density functions stating segmentation lateral and vertical errors and the projection lateral and vertical errors characterization.

As a result, the bi-dimensional probability density function of the CPAs can be derived from previous equation as:

$$f_a(y_{1p}, z_{1p}) = \int_{S_{PCF}} f_1(y'_{1p} - y_{1p'}, z'_{1p} - z_{1p}) f_2(-y'_{1p} - z'_{1p}) dy'_1 dz'_1 \qquad (4)$$

Where:

- f$_a$ is the bi-dimensional probability density function of the CPAs,
- y$_{1p}$ is the minimum predicted lateral separation at the CPA,
- z$_{1p}$ is the minimum predicted vertical separation at the CPA ,
- S$_{PCF}$ is the conflict area
- f$_2$(y$_1$,z$_1$) is the probability density function, representing the distribution of y$_{1p}$ and z$_{1p}$ coordinates errors due to the errors in the segmentation process, and
- f$_1$(y'$_{1p}$,z'$_{1p}$) is the statistically determined bi-dimensional probability density function (pdf) of the CPA'$_P$ coordinates (y'$_{1p}$,z'$_{1p}$) for each projected segment associated to an individual encounter.

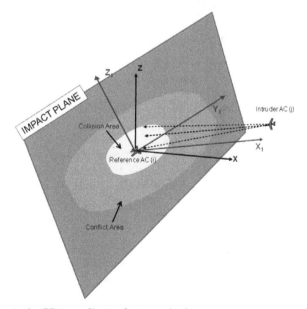

Figure 7. Changes in the CPA coordinates due to projecting errors..

Both expressions estimate the probability of potential collision, having a potential separation violation (potential conflict), for each aircraft encounter, provided that uncertainties in the projection of segmented trajectories and in the segmentation process have been characterised by associated pdfs, f1 and f2, respectively.

4. Results and discussion

The previous mathematical formulation is supported by the previously mentioned ad-hoc software, which has been developed by the authors for Eurocontrol in the framework of the 3D-CRM programme. This software is intended to measure the collision risk in high density ATC en route airspace, based on an analysis of the stored aircraft tracks that have flown in it within a given time frame.

With the purpose of evaluating the mathematical expressions to estimate the probability of collisions, the previously mentioned software tool has been applied to a radar data sample from the Maastricht Upper Area Control Centre (MUAC). EUROCONTROL's Maastricht Upper Area Control Centre (MUAC) is a regional air traffic control centre providing seamless air navigation services in the upper airspace (above 24,500ft) for a large (approximately 700,000 square kilometres) multinational airspace in Europe. An advanced and complex ATC automated system named MADAP (Maastricht Automated Data Processing and Display System) is the technical enabler responsible for managing, processing and presenting in real time information relating to the air traffic flows in the whole area. MADAP performs centralized multi-radar tracking using the information provided by a large number of radars and computes a high quality air traffic situation. In MUAC, a unique horizontal separation standard of 5 NM is used throughout the total area of responsibility. The vertical separation minimum of 1000 ft. is used.

4.1. Empirical estimation for Pa

The general expression of expected Pa is calculated numerically from the relative frequency of potential collisions among all potential conflicts using the following equation:

$$E\left[P(pot.coll/pot.conf)\right] = E\left[P_a\right] \approx \frac{Num.\,of\,pot.\,collisions}{Num.\,of\,pot.\,conflicts} = \frac{19}{35166} = 5.4 * 10^{-4} \qquad (5)$$

Figure 8 illustrates the obtained bi-dimensional histogram of the projected horizontal and vertical separations at the CPA for the whole data period analysed. As it is shown, the number of potential conflicts are higher when encounters are between aircraft established at the same flight level (0ft vertical separation) and, as well, between aircraft having 2.5 and 5NM of lateral separation. It could also be noticed that the number of encounters having 1000ft separation is higher than for any other vertical separation except the 0ft. This is easily understood when taking into account that within the en-route airspace most of the time aircraft are in level flight (namaly always 1000ft apart between contiguous flight levels). If safety barriers have not been applied the number of collisions to happen would have been 19. The area used to compute the number of potential collisions is shown circled by a red dotted circle.

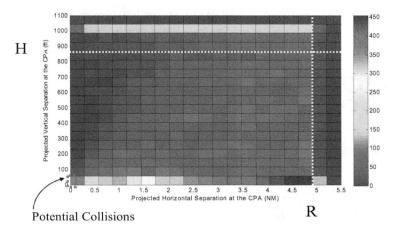

H

Potential Collisions   R

Figure 8. 2D histogram of projected horizontal and vertical separations at the CPA (31 days of radar data)

4.2. Pa estimation for each aircraft encounter

Once the empirical general or expected value for Pa has been obtained, Pa was estimated for each particular encounter by the next expression.

$$
P_a\left(y_p, z_p\right) = 4\lambda_{xy}\lambda_z \cdot \frac{v_x}{\sqrt{v_x^2 + v_z^2}}\left[1 + \frac{\pi}{4}\cdot\frac{\lambda_{xy}}{\lambda_z}\cdot\frac{v_z}{v_x}\right]\cdot f_2{}^*\left(-y_p, -z_p\right) =
$$
$$
= 2\lambda_{xy}\cdot f_{2y}\left(-y_p\right)\cdot 2\lambda_z\cdot\lambda_{2z}\left(-z_p\right)\cdot\frac{v_x}{\sqrt{v_x^2 + v_z^2}}\left[1 + \frac{\pi}{4}\cdot\frac{\lambda_{xy}}{\lambda_z}\cdot\frac{v_z}{v_x}\right]
$$

(6)

This equation provides and individual probability of collision based on:

* kinematics of the encounter (ratio vz,to,vx),
* the predicted minimum lateral separation at the CPA (yp), and
* the predicted minimum vertical separation at the CPA (zp).

It also takes into consideration the segmentation of lateral and vertical errors (f_{2y} and f_{2z}).

A result for Pa estimation for leveled flight encounter is shown in the upper part of figure 8. In this case when CPAp coordinates (yp,zp) are very close to the reference aircraft (ACi), Pa estimated value reaches 3*10-2. This value has a magnitude of two orders higher than the empirical expected result (5.4·10⁻⁴), but strongly decreases when predicted CPAp lays apart from ACi, resulting in values much lower than the empirical one. In the lower part of this figure, the graphs show when one or both aircraft are climbing/descending but having vz/vr ratio close to zero, it could be seen that regardeless the decrease of the maximum value of Pa (7*10-3), It is still greater than the empirical expected result for Pa. Furthermore, the probability of collision for CPAp for which yp coordinates close to zero but zp coordinates separated from the ACi remains significant. Pa estimation for encounters having two different aircraft climbing/descending (vz / vx) ratios is shown in figure 9.

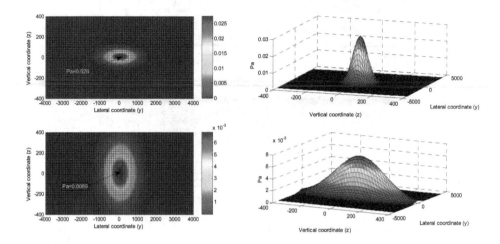

Figure 9. Pa estimation for different CPAp . Aircraft established at a defined flight level or vz equals to zero (upper) and aircraft with vz close to zero (lower).

Despite the fact that the shape of both functions for Pa are similar to the one obtained in the lower part of figure 10 (aircraft climbing/descending and vz/vx close to zero), the maximum values for Pa are different in both cases (9*10-3 for vz/vx=0.1, and 2*10-2 for vz/vx=20), showing that Pa maximum values for CPAp close to reference aircraft (ACi) has a decreasing trend when vz/vr ratio increases. The following table summarises the results obtained from empirical and estimated Pa for the worst case, that is to say Pa for predicted CPAp=(0,0).

The results clearly shows that it is unrealistic to assign the same probability for potential collisions to all potential conflicts, independently of the predicted coordinates for CPA, no matter how these coordinates have been derived.

Empirical result for expected P_a, $E[P_a]$	$5.4 \cdot 10^{-4}$
Estimated P_a for $CPA_p=(0,0)$ and level flight	$3 \cdot 10^{-2}$
Estimated P_a for $CPA_p=(0,0)$ and $v_z/v_x \approx 0$	$7 \cdot 10^{-3}$
Estimated P_a for $CPA_p=(0,0)$ and $v_z/v_x =0.1$	$9 \cdot 10^{-3}$
Estimated P_a for $CPA_p=(0,0)$ and $v_z/v_x =20$	$2 \cdot 10^{-2}$

Table 2. Worst case Pa estimation

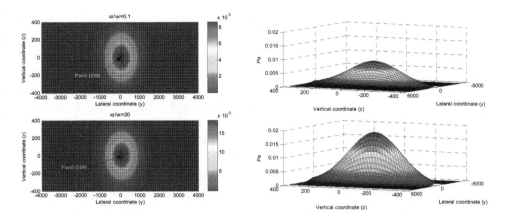

Figure 10. Pa estimation for different CPAp . Aircraft climbing/descending and different vz / vx ratios. vz / vx=0.1(upper), vz / vx=20 (lower)

4.3. Expected Pa estimation for a given scenario and traffic sample

When a collision risk analysis is applied to a representative aircraft population, using segmentation of their stored radar tracks, a 2D histogram of projected horizontal and vertical separations at the CPA can be obtained, as it is shown in figure 8. This histogram provides a first approach for expected Pa using equation (4), which is the way we used to obtain E[Pa]= 5.4*10-4, (this value can taken as reference value for Pa) . If the histogram exhibits a close to uniform distribution, it can be understood that any "generic" potential conflict would became a potential collision with the same probability. It is also possible to propose a different approach to establish the expected value for Pa in a given scenario and for a given aircraft population, discussed below.

$$E[P_a] = \frac{1}{N}\sum_{ji}P_a\left(y_{ji},z_{ji},r_{ji}\right) = \frac{\lambda_{xy}\lambda_z}{N}\sum_{ji}\left[1+\frac{\pi}{4}\cdot\frac{\lambda_{xy}}{\lambda_z}\cdot r_{ji}\right]f_{2y}(y_{ji})\cdot f_{2z}\left(z_{ji}\right) =$$
$$= \frac{\lambda_{xy}\lambda_z}{N}\sum_{ji}\left[1+\frac{\pi}{4}\cdot\frac{\lambda_{xy}}{\lambda_z}\cdot r_{ji}\right]f_{2y}(y_{ji})\cdot f_{2zji}\left(z_{ji}\right)$$

(7)

Where Pa(y$_{ji}$,z$_{ji}$,r$_{ji}$) is the individual probability of each potential collision where:

- rji=vz/vx the between vertical and horizontal relative speeds,
- f2zji the probability density function applied to each aircraft encounter (between each pair of aircraft, i and j).

When this equation is applied to previous MUAC data sample, expected value for Pa results 8.2*10-4, which is slightly higher than the empirical results.

5. Conclusions

This chapter analyse in detail the inherent collision risk involved for each aircraft proximity event by assessing the conditional probability Pa of a potential collision between aircraft that are exposed to risk, that is to say, they are potentially going to violate the separation standards defined for a specific airspace if no corrective action is taken. The proposed approach allows the determination of the severity of each aircraft encounter as the probability of potential collision for each individual aircraft encounter in high density ATC en route airspace, based on an analysis of the stored aircraft tracks that have flown within a given time frame. The authors propose a mathematical formulation to characterise the severity of each aircraft proximity event using the convolution of the bi-dimensional probability density function of the predicted Closest Point of Approach between the aircraft involved and the distribution of lateral and vertical error in the projected position of the aircraft.The presented work aims to provide an individual probability of collision based on the geometry and kinematics of the encounter and the minimum lateral separation and the minimum vertical separation at the predicted Closest Point of Approach or CPA. The formula takes into consideration uncertainties introduced by the radar data error and the segmentation error. The results of this chapter shows that there is not the same severity for all the proximity events on which aircraft pass closer than the prescribed horizontal and vertical separation minima, and also that the expected severity for given a scenario and traffic sample can also vary depending on the kinematic characteristics of aircraft involved within this scenario. It is also considered that collision risk for high density of air traffic can be analysed from the estimation of three different factors:

- Relative frequency of exposition to risk (FeR). The value of this factor can be easily obtained from any radar data sample and strongly depends on the minimum applied horizontal and vertical separations standard and increases with air traffic density,
- Expected severity E(Pa). This value can be directly derived from individual probabilities of potential collision (Pa). Furthermore, having individual severities, it also permits additional assessment on safety (hot spots identification, etc.).
- Expected probability of failure of safety barriers (ATC, TCAS, etc.)

As the two first factors can be derived from the stored tracks of the traffic sample, using the software tool developed by the authors [38], further work is now devoted to develop the probability of failure of the ATM safety barriers. Once the probability of failure were stated and validated, it will be possible to estimate the collision risk for individual encounters, scenarios and air traffic samples. Results obtained for MUAC, with data sample used in previous discussion, exhibits a rounded value for frequency of exposition to risk of FeR=0.3. Probability of potential collision among encounters exposed to risk, Pa or its expected value E(Pa) for the same sample, oscillates between $8.2*10-4$(expected) and $2*10-2$(worst case). Previous results demand a probability of "safety barrier failure" lower than $0.4*10-5$ and $1.7*10-7$ respectively, to reach the ATM en route target level of safety of TLS=10-9. This last value is normally the one used as TLS. For instance, in reference (Eurocontrol, 2006) mid-air collision given as accident frequency (per flight) is 5.4*10-09, specifying that, among them, the frequency of fatal accident, directly caused by ATC (per flight), is 3.5*10-09.

Author details

R. Arnaldo, F.J. Sáez, E. Garcia and Y. Portillo
Universidad Politecnica de Madrid, Madrid, Spain

6. References

[1] Machol, R. E. (1995): "Thirty Years of Modelling Midair Collisions", Interfaces 25: 5 September -October 1995 (151-172)

[2] B. L. Marks (1963) Air traffic control separation standards and collision risk. Royal. Aircraft Establishment Technical Note No. 91, February, 1963

[3] Reich, P.G. (1964), A theory of safe separation standards for Air Traffic Control, Technical Report 64041, Royal Aircraft Establishment, UK

[4] Reich, P. G. (1966). Analysis of Long-range Air Traffic Systems: Separation Standards. Journal of the Institute of Navigation, (19), 88, 169 and 331 (in three parts).

[5] Peter Brooker. Longitudinal Collision Risk for ATC Track Systems: A Hazardous Event Model. Journal of Navigation, 2006, Vol. 59 No. 1. pag. 55-70.

[6] ICAO (1988), Review of the General Concept of Separation Panel, 6th meeting, Doc 9536, Volume 1,ICAO, Montreal, December 1988.

[7] H. D. Sherali C. Smith . Dr. A.A. Trani S. Sale.Q. Chuanwen. Analysis of Aircraft Separations and Collision Risk Modeling. NEXTOR - National Center of Excellence for Aviation Operations Research 1998.

[8] Burt L , October 2000, 3-D Mathematical Model for ECAC Upper Airspace, Final Report

[9] Peter Brooker (Cranfield University).Radar Inaccuracies and Mid-Air Collision Risk: Part 2 En Route Radar Separation Minima The Journal of Navigation (2004).

[10] Carpenter, Brenda D., MIT, Cambridge, MA; Kuchar, James K., MIT, Cambridge, MA. Probability-based collision alerting logic for closely-spaced parallel approach. AIAA-1997-222 Aerospace Sciences Meeting and Exhibit, 35th, Reno, NV, Jan. 6-9. 1997.

[11] Kuchar, James K., MIT, Cambridge, MA; Winder, Lee F., MIT, Cambridge. Generalized philosophy of alerting with applications to parallel approach collision prevention. MA AIAA-2001-4052 AIAA Guidance, Navigation, and Control Conference and Exhibit, Montreal, Canada, Aug. 6-9, 2001.

[12] Lee F. Winder, ; James K. Kuchar. Evaluation of Collision Avoidance Maneuvers for Parallel Approach . Journal of Guidance, Control, and Dynamics 0731-5090 vol.22 no.6 (801-807)doi: 10.2514/2.4481, 1999

[13] Powell, J. David, Stanford Univ., CA; Houck, Sharon, Stanford Univ., CA Assessment of the possibility of a midair collision during an ultra closely spaced parallel approach. . AIAA-2001-4205 AIAA Guidance, Navigation, and Control Conference and Exhibit, Montreal, Canada, Aug. 6-9, 2001.

[14] Sharon W. Houck, J. David Powell, Probability of Midair Collision During Ultra Closely Spaced ParallelApproaches. Journal of Guidance, Control, and Dynamics, 0731-5090 vol.26 no.5 (702-710) do: 10.2514/2.5124. 2003.

[15] Shepherd, Roger, Rannoch Corp., Alexandria, VA; Cassell, Rick. A reduced aircraft separation risk assessment model, VA AIAA-1997-3735 AIAA Guidance, Navigation, and Control Conference, New Orleans, LA, Aug. 11-13, Collection of Technical Papers. Pt. 3 (A97-37001 10-63). 1997.

[16] Blom, H.A.P., Bakker, G.J., Blanker, P.J.G., Daams, J., Everdij, M.H.C., and Klompstra, M.B. "Accident Risk Assessment for Advanced ATM," In: Air Transportation Systems Engineering, G.L. Donohue and A.G. Zellweger (Eds.), AIAA, 2001, pp. 463-480. 2001.

[17] H. Blom, GJ Bakker, B. Klein Obbink and MB Klompstra. Free Flight safety risk modeling and simulation. Proceedings of 2nd International Conference on Research in Air Transportation ICRAT 2006, at Beograd, Serbia, June 24-28, 2006.

[18] H. Blom,; B. Klein Obbink, B. Bakker, National Aerospace Laboratory NLR. Safety Risk Simulation of an Airborne Self Separation Concept of Operation. AIAA-2007-7729 7th AIAA ATIO Conf, 2nd CEIAT Int'l Conf on Innov and Integr in Aero Sciences,17th LTA Systems Tech Conf; followed by 2nd TEOS Forum, Belfast, Northern Ireland, Sep. 18-20, 2007.

[19] Peter Brooker. Air Traffic Management accident risk. Part 1: The limits of realistic modeling. Safety Science 44, 419–450. 2006.

[20] Paielli, R.A. and H. Erzberger, "Conflict probability estimation for free flight", AIAA J. of Guidance, Control and Dynamics, Vol. 20, pp.588-596. 1997.

[21] Paielli, R.A. and H. Erzberger, "Conflict Probability Estimation Generalised to Non-Level Flight", Air Traffic Control Quarterly, Vol. 7, pp.195-222, 1999.

[22] Prandini, M., J. Hu, J. Lygeros and S. Sastry, A probabilistic approach to aircraft conflict detection, IEEE Tr. on Intelligent Transportation Systems, Vol. 1, No. 4, pp. 199-220. 2000.

[23] M. Prandini, J. Lygeros, A. Nilim, and S. Sastry, "A Probabilistic Framework for Aircraft Conflict Detection", AIAA-99-4144, in Proc. AIAA Guidance, Navigation, and Control Conf., Portland, OR, August 9-11, pp. 1047-1057.1999.

[24] ICAO, Annex 11 – Air Traffic Services, 12th edition, incorporating amendments 1-38, July 1998, Green pages, attachment B, paragraph 3.2.1.

[25] Kuchar, J. and Yang, L., "A Review of Conflict Detection and Resolution Modelling Methods," IEEE Transactions on Intelligent Transportation Systems, Vol. 1, No. 4, pp. 179–189. December 2000.

[26] Kuchar and L. Yang, "Survey of Conflict Detection and Resolution Modelling Methods", AIAA-97-3732, in Proc. AIAA Guidance, Navigation, and Control Conf., New Orleans, LA, August 11-13, 1997.

[27] Henk A.P. Blom , Bart Klein Obbink, G.J. (Bert) Bakker. Safety risk simulation of an airborne self separation concept of operation, 7th AIAA Aviation Technology, Integration and Operations Conference (ATIO)
2nd C 18 - 20, Belfast, Northern Ireland AIAA 2007-7729. September 2007.

[28] Lee Yang*, Ji Hyun Yang†, James Kuchar‡, Eric Feron§ Massachusetts Institute of Technology, Cambridge, MA. A Real-Time Monte Carlo Implementation for Computing Probability of Conflict. AIAA Guidance, Navigation, and Control Conference and Exhibit 16 - 19, Providence, Rhode Island. August 2004.

[29] Burt L , October 2000, 3-D Mathematical Model for ECAC Upper Airspace, Final Report.

[30] Campos L. M. B. C. ; Marques J. M. G. On the probability of collision between climbing and descending aircraft , ; Journal of aircraft ISSN 0021-8669 CODEN JAIRAM / vol. 44, no2, pp. 550-557. 2007.

[31] Campos L. M. B. C.. Probability of collision of aircraft with dissimilar position errors , ; Journal of aircraft ISSN 0021-8669 CODEN JAIRAM , vol. 38, no4, pp. 593-599. 2001.

[32] Leonard A. Wojcik The MITRE Corporation, McLean, VA 22102 U.S. Probabilistic Aircraft Conflict Analysis for a Vision of the Future Air Traffic Management System. AIAA 5th Aviation, Technology, Integration, and Operations Conference (ATIO) 26 - 28, Arlington, Virginia. September 2005.

[33] Leonard A.Wojcik* The MITRE Corporation, McLean. Probabilistic Aircraft Conflict Analysis for a Future Air Traffic Management System. VA 22102 DOI: 10.2514/1.22850 Journal of Aerospace Computing, Information, and Communication Vol. 6, June 2009

[34] Rachelle L. Ennis1 and Yiyuan J. Zhao2 University of Minnesota, Mpls, MN. Defining Appropriate Inter-Aircraft Separation Requirements. AIAA 4th Aviation Technology, Integration and Operations (ATIO) Forum 20 - 22, Chicago, Illinois. September 2004.

[35] Jerry Dingy Claire Tomlinz. A Dynamic Programming Approach for Aircraft Conflict Detection. AIAA Guidance, Navigation, and Control Conference, 10 - 13, Chicago, Illinois . August 2009.

[36] D. E. Stepner. Modelling of Aircraft Position Errors with Independent Surveillance. VOL. 11, NO. 9, AIAA Journal 1273. September 1973.

[27] Eduardo José García González, INECO, Madrid, Spain, Francisco Javier Sáez Nieto, Polytechnic University of Madrid, Madrid, Spain, Maria Isabel Izquierdo, EUROCONTROL, Brussels, Bélgica . Identification and analysis of proximate events in high density en route airspaces Paper N° 63]. 7th USA – EUROPE ATM R&D Seminar July 02-05, Barcelona. 2007

[38] Saez F, Arnaldo R, Garcia E, McAuley G, Izquierdo M. Development of a three dimensional collision risk model tool to asses safety in high density en-route airspaces. DOI:10.1243/09544100JAERO704

[39] Jaroslav Krystul.Modelling of stochastic hybrid systems with applications to accident risk assessment. 6 September 2006

[40] Rachelle L. Ennis† and Yiyuan J. Zhao‡ University of Minnesota, Minneapolis, Minnesota. Characterization of Aircraft Protected Zones. AIAA's 3rd Annual Aviation Technology, Integration, and Operations (ATIO) Tech 17 – 19, Denver, Colorado. November 2003,

[41] J. F. Bellantoni. The Calculation of Aircraft Collision Probabilities DOT-TSC-FAA-71-27, October'1971.

[42] Reynolds T.G., Hansman R.J., Analysis of Aircraft Separation Minima Using a Surveillance State Vector Approach, MIT, 2001

[43] Ennis, R. L.; Zhao,Y.J., Defining Appropriate Inter-Aircraft Separation Requirements, AIAA's 4th Annual Aviation Technology Integrations and Operations (ATIO) Forum # AIAA2004-6203, September 20-22, 2004

Chapter 7

How to Manage Failures in Air Traffic Control Software Systems

Luca Montanari, Roberto Baldoni, Fabrizio Morciano,
Marco Rizzuto and Francesca Matarese

Additional information is available at the end of the chapter

1. Introduction

Failure Management consists of a set of functions that enable the detection, isolation, and correction of anomalous behavior in a monitored system trying to prevent system failures. An effective failure management should monitor the system looking for errors and faults that could end up in a failure and overcome such issues when they arise.

Air Traffic Control (ATC) systems are large and complex systems supervising the aircraft trajectories from departure to destination. Such systems have hard reliability and dependability requirements. Having an effective failure management in such kind of critical systems is a must for safety and security reasons. Two main approaches have been developed in the literature to implement these failure management systems:

- Reactive Fault Management;
- Proactive Fault Management.

Due to the complexity and the strong requirements, current ATC systems adopt both of them. The Reactive Fault Management is based on the detection paradigm. A reactive fault manager get triggered at the moment in which errors occur and should have the following capability: diagnosis, symptom monitoring, correlation, testing, automated recovery, notification, online system topology update. The Proactive Fault management scheme anticipates the formation of erroneous system states before it actually materializes into a failure. Known techniques in this field are rejuvenation of system components [24], checkpointing [4], prediction mechanisms [21]: to predict a failure occurrence and thus triggering the system state recovery.

This chapter focus on failure management in ATC systems pointing out motivations that led engineers to do specific design choices. Two case studies as real implementations of the paradigms are also presented: a reactive approach deployed in a real ATC System and a novel proactive approach that has the distinctive features to be (i) *black-box*: no knowledge of applications' internals and logic of the mission critical distributed system is required (ii) *non-intrusive*: no status information of the nodes (e.g., CPU) is used; and (iii) *online*: the analysis is performed during the functioning of the monitored systems.

The chapter is organized as follows: section 2 explains the motivations of failure management in ATC. Section 3 shows the role that faults and failures have in ATC systems and the relationship with safety regulation. Section 4 presents the objectives of failure management while section 5 investigates proactive, reactive approaches and we will introduce the online failure prediction technique. The sections 6 and 7 present the two case studies, one using a reactive approach and one that use a proactive approach. Section 8 concludes the chapter.

2. Motivations

Distributed mission critical systems such ATC, battlefield or naval command and control systems consist of several applications distributed over a number of nodes connected through a LAN or WAN. The applications are constructed out of communicating software components that are deployed on those nodes and may change over time. The dynamic nature of applications is mainly due to (i) adopted policies to cope with software or hardware failures, (ii) load balancing strategies and (iii) the management of new software components joining the system. Additionally such systems have to react to input in a soft real time way, i.e., an output has to be provided after a few seconds from the input the generated it. In such complex real time systems, failures may happen with potentially catastrophic consequences for their entire functioning. The industrial trend is to face failures by using appropriate software engineering techniques at the design phase. However these techniques cannot reduce to zero the probability of failures during the operational phase due to the unpredictability and uncertainty behind a distributed systems [7], thus there is the need of supervising services that are not only capable of detecting a failure, but also predicting and preventing it through an analysis of the overall system behavior.

The literature about failures and fault management embraces several aspects: reactive approaches, proactive approaches, fault detection, failure detection, faults and failure isolation, failure prediction. Before investigating some details of these techniques, it is important to point out some definitions that will be used along this chapter [11]:

- The *system behavior* is what the system does to implement its function;
- A *failure* is an event that occurs when the delivered service deviates from correct service;
- A *fault* is the cause of an error.
- An *error* is a deviation in the sequence of the system's states.

A fault is *active* when it produces an error, it is *dormant* otherwise. The next section specializes the faults and failures in the ATC domain.

3. Faults and failures in ATC systems and relationship with safety regulation 482/2008

An ATC system is a large and complex system with several interrelated functions. It receives inputs from several heterogeneous actors like: messages from external lines (e.g. AFTN), radar information, radio communications with aircraft etc. All these information need to be integrated, processed, correlated and finally presented to an ATC system as a global operational picture of the sky. A controller looks at this picture and, according to the adopted procedures, addresses the aircraft pilot in the safest way ensuring to select the most efficient trajectory for reaching the final destination.

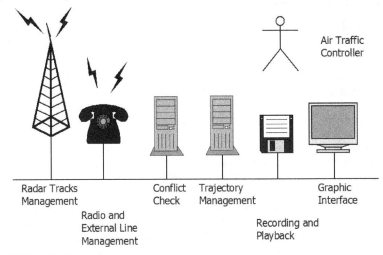

Figure 1. ATC Very High Level Architecture

A very high level architecture of an ATCs is shown in Fig. 1. The Figure highlights the needs for an ATCs in term of hardware, software and human factors. The number of components involved can change depending on the vendor, size of the system and requirements from the customers, still to give a rough idea of the order of magnitude of the size of the system, an ATC system is several million lines of code.

ATCs does not require strict real-time time of responses (the separation criteria can be around seconds) but the availability of the system should be greater than 99,99%. An ATCs architecture requires at least the following capabilities:

- discovering a fault in a predictable time;
- sharing the same data among all the components forming the system;
- maintaining the service or restore it in a predictable time.

According to the previous criteria we can identify some class of faults:

- misalignment in time (not all the system is aware of its processing capacity);
- misalignment in data (not all the system shares the same information);
- misalignment in functionalities (not all the capabilities are available).

The first class of faults implies failures related to delay in communications (inside and outside the system) and, human factor (wrong order). Faults related to the hardware are minimized by a proper configuration and tuning of the ATC system. In the worst case the entire ATC system can be replaced by a different one using a separate network and possibly employing different hardware and software components to exploit diversity argument (sometimes the previous version of the ATC system is used as fallback).

The second class of faults implies failures related to the mismatch between the output of processing server in the system; part of the system could process data no longer relevant with respect to the real status of ATC system. This impacts ATC systems as they cannot rely anymore on the information provided by the system.

The third class of faults implies failures related to degraded system usability, part of the system cannot be used and its functionality cannot be accessed by ATC systems or software components.

Safety is an essential characteristic of AirTrafficManagement/ATC functional systems. It has a dominant impact upon operational effectiveness. ATM/ATC functional systems in now evolving in a continuously growing integrated environment including automation of operational functions, formerly performed through manual procedures and massive and systematic use of software. All this has a prominent impact for the achievement of safety [5]. Moreover, regulatory compliance has become a legal and necessary extension of business continuity with an increasingly complex set of laws and regulations relating to data integrity and availability. Ensuring the integrity and availability for ATC systems bring bad and good news on regulatory compliance. The bad news is the regulations do not provide a "blueprint" for protection. The good news is high availability and continuous availability protection strategies will help you meet these regulatory requirements, minimizing the risk that under-protected systems will create breaks in the "chain of data". It is important to note that compliance is a moving target; both government and industry leaders will continue to move toward more specific regulations and standards [12]. The issue of regulatory compliance has became more acute on 1st January 2009 when the Regulation (EC) 482/2008 "establishing a software safety assurance system to be implemented by air navigation service providers" went into effect [3]. Still, laws or regulations do not set a specific process or specific requirements for an ATC system, they just describe expected outcomes. The Software Fault Management System supports business continuity and Regulation (EC) 482/2008 compliance, by identifying a set of "risk-mitigation means", defined from the risk-mitigation strategy achieving a particular safety objective. Moreover, it provides:

- "cutover or hot swapping", that is the approach of replacing European air traffic management network (EATMN) system components while the system is operational;
- "software robustness", that is the robusteness of the software in the event of unexpected inputs, hardware faults and power supply interruptions, either in the computer system itself or in connected devices; and
- "overload tolerance", that is the tolerance of the system to, inputs occurring at a greater rate than expected during normal operation of the system.

4. Failure management objectives

The objective of the failure management can be broadly divided in *Failure Detection, Failure Isolation, Failure Identification*. Failure detection and isolation has become a critical issue in the operation of high-performance ships, submarines, airplanes, space vehicles, and structures, where safety, mission satisfaction, and significant material value are at stake. [8] presents a survey on the failure detection techniques introducing some basic definition:

1. *Failure Detection* is the task to produce an indication that something is going wrong, i.e. a failure is present in the system.

2. *Failure Isolation* is the determination of the exact location of a failure.

3. *Failure Identification* is the determination of the size of the failure.

While detection and isolation are a must in any mission critical system, failure identification can be an overkill and then sometime it is not implemented.

5. Failure management techniques - reactive and proactive

5.1. Reactive approach

The reactive approach in fault management is based on the detection paradigm. A reactive fault manager gets triggered at the moment in which errors occurs. More specifically, in order to achieve it's main goals, it is necessary to have the following capabilities [10]:

- *Symptom monitoring*: Symptoms are manifestations of underlying faults and must be monitored to detect the occurrence of problems as soon as they happen. A fault manager quality is its response time to symptoms. The quicker this reaction occurs, the higher the probability to recover the system error is. This in turn raise the probability that the fault will not end up into a failure. In our case study the middleware platform used makes use of FT-CORBA (Fault tolerant CORBA) which rely on the fault detection to implement tolerance logics (see section 6)

- *Diagnosis*: identifies the root causes of "known" symptoms. A fault may originate on one component and then it could manifest on some other component. In large scale systems, there is no one-to-one mapping between faults, errors, failures. Studies on such systems have shown that typically up to 80% of the fault management effort is spent in identifying root causes after the manifestation of symptoms [23].

- *Correlation*: a correlation capability provides knowledge about root causes of "known" symptoms to the diagnosis modules. Modern systems are often richly instrumented with a large number of sensors that provide large amounts of information in the form of messages and alarms. This flow of information cannot be handled by humans in real-time as a small number of roots causes results in a huge number of messages and alarms. Therefore it is necessary to provide them with concise and aggregate notifications of underlying root causes. Correlation is the process of recognizing and organizing groups of events that are related each other.

- *Testing*: in large software systems, it is impractical (and sometime impossible) to monitor every variable. Instead key observable variables are monitored to generate symptom events. Diagnostic inference typically identifies a set of suspected root causes. A test planning facility is needed to select additional variables to be examined to isolate the root causes. The fault management application then needs to request or run these tests, and utilizes their results to complete the diagnosis. A test, as originally defined in [22], can incorporate arbitrarily complex analysis and actions, as long as it returns a true or false value.

- *Automated recovery*: identifying and automating recovery procedures facilitate rapid response to problems and allow for growth in equipment, processes, and services, without increasing the supervisory burden on system operators. The automation in recovery decreases the response time to an error and thus decreasing the probability that it may cause a proper system failure.

- *Notification*: system operators require notifications of all critical fault management activity, especially the identification of root causes, and causal explanations for alarms, tests, and repair actions in a manner that they can follow easily. Sometimes they need to distinguish between what is observed by system sensors versus what is inferred by the underlying fault management application.

- *Postmortem*: information from diagnostic problem solving is fed back to the fault management system for historic record keeping in order providing enough data for offline failure analysis to discover some of the mappings between failures and their root cause. It is important to underline that this analysis is different than the offline analysis to discover failure patterns. Failure patterns and relationships between failures and the root cause are orthogonal concepts even if some relationships between failures and faults can form a failure pattern. That's because failures are not caused just by errors or faults but also by system configurations and human interaction patterns.

- *Online system topology update*: in an ATC system the reactive fault manager should support expert systems for effective diagnosis of root causes of system errors and that the expert system uses a knowledge base to infer the right diagnosis. The knowledge base as a module can be replaced or connected to another knowledge base. Other components can be completely removed or added. All this dynamic changes need to be done at run-time. It may not be feasible indeed to take the fault management system off-line each time that there is a change in the system topology.

5.2. Proactive fault management

Using reactive schemes there are limits to increasing mission critical systems availability. Failure management started looking at proactive approaches to overcome these limitations such as rejuvenation of system components [10]. This scheme anticipates the formation of erroneous system states before it actually materializes into a failure. The listed schemes to increase system availability can be a more effective idea if applied intelligently and preventively. The question remains: when we should apply check-pointing and rejuvenation techniques? To answer this question we need a way to tell if the current state of the system is going to evolve into a failure state. We can extend this concept to include parts or the entire history of the systems state transitions. So to answer the question of the ideal trigger timing for high availability schemes we need to develop a model of the system in question which allows us to derive optimized triggering times. To increase availability of a software system during runtime basically two main concepts are involved: The method to re-initiate the system or a component to a failure free state like rejuvenation and a prediction mechanism to predict a failure occurrence and thus trigger the system state recovery.

5.3. Online failure prediction

The problem of modeling a system had always the main objective of predicting its behavior. A significant body of work has been published in this area. As far as distributed systems is concerned, a recent work [20] introduces a taxonomy that structures the manifold of approaches. The more relevant approaches for the ATC purpose are the following ones:

- Symptoms Monitoring: Manifestation of faults is not necessary a clear situation rather than a more fuzzy one. It can influence the hosted system gradually in time and space. This type of symptoms is called service degradation. A prominent example for such types of symptoms in ATC or, in general, in mission critical systems is *response-time*. The fault underlying this symptom might be a bad process priority management and consequent starvation of some other processes having lower priority. The key notion of failure prediction based on monitoring data is that faults like starvation in priority management can be detected by their side effects such as high response-time. These side-effects are called symptoms. Later on this chapter (section 7) we will show an architecture for online failure prediction in ATC that uses symptoms monitoring.

- Error Detection: Once a fault manifests itself, it becomes an error. Errors and symptoms are different: symptoms are the observation of system state over time; a symptom is a behavior that deviates from the "normal" behavior. While error is something that actually goes wrong. The fault at this stage did not develop in service failure yet but it would possibly do it. What is the probability that this error ends up in a failure? for how long since the first occurrence this probability keeps high? Error detection approaches attempt to answer these questions. The error detection usually employs online failure predictors based on rules, data-mining approaches, pattern recognition, fault trees etc.

6. Reactive approach case study: FT CORBA in a real ATC system

6.1. Motivation

"There shall be no single point of failure", this is one of the basic requirements for any ATC system. It drives alone many choices about the design, the used technologies, the verification strategies of a complex distributed system which has to provide a very high service availability : at least 99,99% i.e. downtime of about 5 minutes per month. The complexity of such systems is more and more stored in the software, which is error prone to problems injected at design or coding time as well as to unexpected scenarios due to runtime concurrency and other factors, like for example upgrading activities. Then software fault tolerance stands beside the traditional hardware based solutions and often replaces them, considering also that these systems are maintained and can evolve over a 25 years lifecycle: any chosen solution must support changes. In this context FT CORBA is widely used in ATC, but also in Naval Combat Management and other Command and Control systems. FT CORBA provides both replication and failure transparencies to the application and moreover it is standardized by the Object Management Group [18] [15].

6.2. Principle of FT CORBA

The FT CORBA specification defines an architecture and a framework for resilient, highly-available, distributed software systems suitable for a wide range of applications, from business enterprise applications to distributed, embedded, real-time applications. The basic concepts of FT CORBA are entity redundancy, fault detection and fault recovery; replicated entities are several instances of CORBA objects that implement a common interface and thus are referenced by an object group (Interoperable Object Group Reference, IOGR). IOGRs lifecycle and update are totally managed by the FT CORBA infrastructure; client applications are unaware of object replication and changes in the object group due to replica failure are transparent since their request are forwarded to the right replica. The infrastructure (see Fig. 2) provides means to monitor the replicated objects and to communicate the faults, as well as to notify the fault to other interested parties, which could contribute to recover the application. Beyond replication, object groups and complete transparency, FT CORBA relies

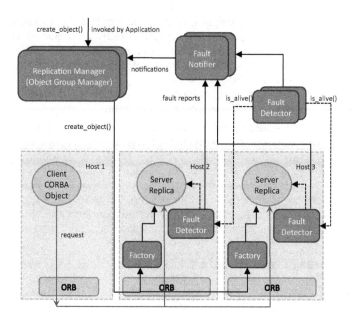

Figure 2. FT CORBA framework

also on infrastructure-controlled consistency. Strong replica consistency is enforced in order to guaranty that the sequence of requests invoked on the object group passes unaltered across the fault of one or more replicas.

6.3. Specialization of FT CORBA for safety critical systems: CARDAMOM use case

In the following we are going to focus on the design choices made for a significant piece of a real ATC system, namely CARDAMOM [2], and that is implemented in a CORBA based middleware.

Among the different replication styles, CARDAMOM adopts the warm passive approach to replicate statefull servers: during normal operation, only one member of the object group, the primary replica, executes the methods invoked on the group. The backup replicas are warm because they receive the status updates at the end of each request from the primary; this way they are always ready to process the next request, in case the primary fails. The FT infrastructure is in charge of detecting such failure and of triggering the switch to a new primary. Transferring to the backup replicas the updated status and the list of processed request ids, it is guaranteed that requests are always served exactly once as long as there are available replicas.

The software architecture is based on CORBA Component Model (see Fig. 3) and then the natural unit of redundancy is a component of the CCM; This Component is a unit of design, development and deployment realized through a collection of CORBA Objects which define attributes and interfaces, called ports [14]. In this context, the exposed ports (facets) of the server components are defined as objects of FT CORBA groups. This approach suits

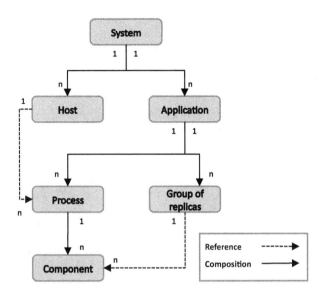

Figure 3. System decomposition in application, process, component, group and host.

well with FT CORBA specification but put in evidence an operative need: in Operating Systems that manage the process as unit of memory space and failure (e.g. POSIX process in Linux/Unix), monitoring and recovery should be done at process level. Then CARDAMOM restricts FT CORBA entity redundancy by enforcing that within the same process all replicated components play the same role, that is all primaries or all backups. This need is also tackled by an extension of FT CORBA specification, the beta OMG specification "Lightweight Fault Tolerance for Distributed RT Systems" [17].

A very important aspect of CARDAMOM is the fault detection; since the framework is tuned to react and recover from failures, namely a process crash, mechanisms are put in place to detect malfunctions like for example deadlocks or endless loops which do not lead necessarily to a crash. After the detection, most of the times the safest action to recover normal behavior is to stop or kill the faulty process in order to trigger a switch to a new replica. Normally fault detectors work with several patterns at the same time: they can use a pull model, e.g. "is alive" call, or push model, e.g. by handling OS signals to detect the death of processes or even be signaled by the application itself after a fatal error.

FT CORBA with warm-passive replication style fits well the need of statefull servers which must guarantee the processing of sequenced requests. However, an ATC system needs other components to be resilient to failures act as stateless components. Generally speaking, stateless components have to provide their services with high availability but do not need to check for "exactly once" semantics of client requests either to support the state transfer. In this case it is used the Load Balancing framework, specified at OMG by the Lightweight Load Balancing specification [16]. It reuses the object group definition of FT CORBA and allows to transparently redirect the client requests among a pool of server replicas according to predefined or user defined strategies, for example through random or round-robin policies. In this way two conflicting goals are achieved at the same time: distribute the computational load among several resources and supporting fault tolerance. because fault detectors are used to update the object group in case of failure and activate recovery mechanism. An additional and important feature is also to prevent that several replicas may crash because of the same implementation: by means of fine request identification, the framework allows to stop those requests that have caused failures, thus avoiding repetitive crashes which would result in a complete system failure.

Middleware CARDAMOM provides all the previously mentioned services (see Fig. 4): in fact it has been chosen as the foundation for a safety critical subsystem, the core part of a next generation ATC system.

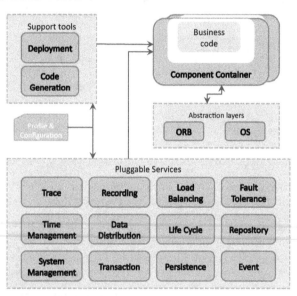

Figure 4. CCM and CORBA based middleware services.

In order to separate duties and define a clearly decoupled architecture that could support extensibility and maintainability, a three tier model has been put in place for the building blocks of the ATC system using CARDAMOM services. The first tier provides the interface to the external clients and guaranties the ordered processing of requests; it is realized by statefull components replicated with FT CORBA and warm passive replication style. The second tier executes the business logic; it is realized by stateless components replicated with LwLB supporting fault containment for killer requests; the third tier tackles the data management and persistency.

This architecture (see Fig. 5) is proven to be, at the same time, resilient to failures and highly scalable in terms of computational power, thus responding to the opposite requirements coming from availability, safety and performances. The use of FT and LB CORBA services is strongly interrelated also with System Management services, that are informed of replica crashes by the Fault Notifier. Automatic actions are put in place in order to stop or restart the replicas and contribute to the overall system availability; actions like restart and stop can be defined with different level of granularity, that is for process, application or host according to

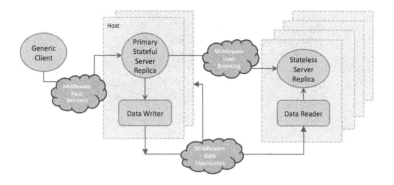

Figure 5. 3-tier architecture.

the kind of failure. As final consideration it is very important to underline that the design and implementation of the middleware services that provide this fault tolerant framework have to be themselves fault tolerant.

7. Failure prediction case study: CASPER

In this section we introduce the design, implementation and experimental evaluation of a novel online, non-intrusive and black-box failure prediction architecture we named CASPER that can be used for monitoring mission critical distributed systems. CASPER is (i) *online*, as the failure prediction is carried out during the normal functioning of the monitored system, (ii) *non-intrusive*, as the failure prediction does not use any kind of information on the status of the nodes (e.g., CPU, memory) of the monitored system; only information concerning the network to which the nodes are connected is exploited as well as that regarding the specific network protocol used by the system to exchange information among the nodes (e.g., SOAP, GIOP); and (iii) *black-box*, as no knowledge of the application's internals and of the application logic of the system is analyzed. Specifically, the aim of CASPER is to recognize any deviation from normal behaviors of the monitored system by analyzing symptoms of failures that might occur in the form of anomalous conditions of specific performance metrics. In doing so, CASPER combines in a novel fashion Complex Event Processing (CEP) [13] and Hidden Markov Models (HMM) [19]. The CEP engine computes at run time the performance metrics. These are then passed to the HMM in order to recognize symptoms of an upcoming failure. Finally, the symptoms are evaluated by a failure prediction module that filters out as many false positives as possible and provides at the same time a failure prediction as early as possible. We deployed CASPER for monitoring a real ATC system. Using the network data of such a system in the presence of both steady state performance behaviors and unstable state behaviors, we first trained CASPER in order to stabilize HMM and tune the failure prediction module. Then we conducted an experimental evaluation of CASPER that aimed to show its effectiveness in timely predicting failures in the presence of memory and I/O stress conditions.

7.1. Failure and prediction model

We model the distributed system to be monitored as a set of nodes that run one or more services. Nodes exchange messages over a communication network. Nodes or services can be

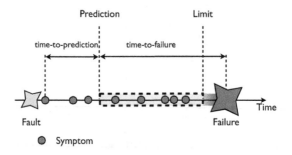

Figure 6. Fault, Symptoms, Failure and Prediction

subject to failures. A failure is an event for which the service delivered by a system deviates from its specification [11]. A failure is always preceded by a fault (e.g., I/O error, memory misusage); however, the vice versa might not be always true. i.e., a fault inside a system could not always bring to a failure as the system could tolerate, for example by design, such fault.

Faults that lead to failures, independently of the fault's root cause, affect the system in an observable and identifiable way. Thus, faults can generate side-effects in the monitored systems till the failure occurs. Our work is based on the assumptions that a fault generates increasingly unstable performance-related symptoms indicating a possible future presence of a failure, and that the system exhibits a steady-state performance behavior with a few variations when a non-faulty situation is observed [25]. In Figure 6 we define *Time-to-failure* the distance in time between the occurrence of the prediction and the software failure event. The prediction has to be raised before a time *Limit*, beyond which the prediction is not sufficiently in advance to take some effective actions before the failure occurs. We also consider the *time-to-prediction* which represents the distance between the occurrence of the first symptom of the failure and the prediction.

7.2. CASPER architecture

The architecture designed is named CASPER and is deployed in the same subnetwork as the distributed system to be monitored. Figure 7 shows the principal modules of CASPER that are described in isolation as follows.

Pre-Processing module. It is mainly responsible for capturing and decoding network data required to recognize symptoms of failures and for producing streams of events. The network data the Pre-Processing module receives as input are properly manipulated. Data manipulation consists in firstly decoding data included in the headers of network packets. The module manages TCP/UDP headers and the headers of the specific inter-process communication protocol used in the monitored system (e.g., SOAP, GIOP, etc) so as to extract from them only the information that is relevant in the detection of specific symptoms (e,g., the timestamp of a request and reply, destination and source IP addresses of two communicating nodes). Finally, the Pre-Processing module adapts the extracted network information in the form of *events* to produce streams for the use by the second CASPER's module (see below).

Symptoms detection module. The streams of events are taken as input by the Symptoms detection module and used to discover specific performance patterns through complex event processing (i.e., event correlations and aggregations). The result of this processing is a system

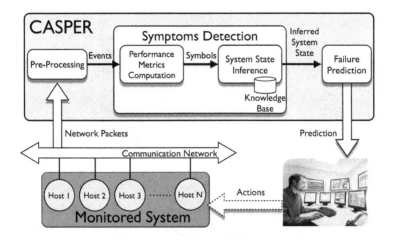

Figure 7. The modules of the CASPER failure prediction architecture

state that must be evaluated in order to detect whether it is a safe or unsafe state. To this end, we divided this module into two different components, namely a *performance metrics computation* component and a *system state inference* component.

The performance metrics computation component uses a CEP engine for correlation and aggregation purposes. It then periodically produces as output a representation of the system behavior in the form of *symbols*. Note that, CASPER requires a *clock mechanism* in order to carry out this activity at each *CASPER clock cycle*. The clock in CASPER allows it to model the system state using a discrete time Markov chain and let the performance metrics computation component coordinate with the system state inference one (see below). The representation of the system behavior at run time is obtained by computing *performance metrics*, i.e., a set of time-changing metrics whose value indicates how the system actually works (an example of network performance metric can be the round trip time). In CASPER we denote symbols as σ_m (see Figure 8), where $m = 1, \ldots, M$. Each symbol is built by the CEP engine starting from a vector of performance metrics: assuming P performance metrics, at the end of the time interval (i.e. the clock period), the CEP engine produces a symbol combining the P values. The combination of performance metrics is the result of a discretization and a normalization: each continuous variable is discretized into slots of equal lengths. The produced symbol represents the state of the system during the clock period.

The system state inference component receives a symbol from the previous component at each CASPER clock cycle and recognizes whether it is a correct or an incorrect behavior of the monitored system. To this end, the component uses the Hidden Markov Models' forward probability [19] to compute the probability that the model is in a given state using a sequence of emitted symbols and a knowledge base(see Figure 7). We model the system state to be monitored by means of the *hidden process*. We define the states of the system (see Figure 8) as *Safe*, i.e., the system behavior is correct as no active fault [11] is present; and *Unsafe*, i.e., some faults, and then symptoms of faults, are present.

Failure Prediction module It is mainly responsible for correlating the information about the state received from the system state inference component of the previous CASPER module. It takes in input the inferred state of the system at each CASPER clock-cycle. The inferred state

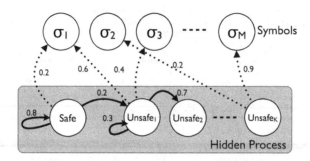

Figure 8. Hidden Markov Models graph used in the system state inference component

can be a safe state or one of the possible unsafe states. Using the CEP engine, this module counts the number of consecutive *unsafe* states and produces a failure prediction alert when that number reaches a tunable threshold (see below). We call this threshold *window size*, a parameter that is strictly related to the *time-to-prediction* shown in Figure 6.

7.2.1. Training of CASPER

The knowledge base concerning the possible safe and unsafe system states of the monitored system is composed by the parameters of the HMM. This knowledge is built during an initial training phase. Specifically, the parameters are adjusted by means of a training phase using the max likelihood state estimators of the HMM [19]. During the training, CASPER is fed concurrently by both recorded network traces and a sequence of pairs `<system-state,time>`. Each pair represents the fact that at time `<time>` the system state changed in `<system-state>`[1].

7.2.2. Tuning of CASPER parameters

CASPER architecture has three parameters to be tuned whose values influence the quality of the whole failure prediction mechanism in terms of false positives and time-to-prediction. These values are (i) the length of the CASPER *clock period*; (ii) the *number of symbols* output by the performance metrics computation module; (iii) the length of the failure prediction, i.e., *window size*.

The length of the clock period influences the performance metrics computation and the system state inference: the shorter the clock period is, the higher the frequency of produced symbols is. A longer clock period allows CASPER to minimize the effects of outliers. The number of symbols influences the system state inference: if a high number of symbols is chosen, a higher precision for each performance metrics can be obtained. The failure prediction window size corresponds to the minimum number of CASPER clock cycles required for raising a prediction alert. The greater the window size, the more the accuracy of the prediction, i.e., the probability that the prediction actually is followed by a failure (i.e. a true positive prediction). The tradeoff is that the time-to-prediction increases linearly with the windows size causing shorter time-to-failure (see Figure 6); During the training phase, CASPER automatically

---

[1] As the training is offline, the sequence of pairs `<system-state,time>` can be created offline by the operator using network traces and system log files.

chooses the best values for both clock period and number of symbols, leaving to the operator the responsibility to select the windows size according to the criticality of the system to be monitored.

7.3. Monitoring a real ATC system with CASPER

CASPER has been tested on the same real ATC system used in the reactive approach case study (section 6). CASPER intercepts GIOP messages produced by the CORBA middleware and extracts several information from them in order to build the representation of the system at run time. In this section we describe how the events are represented starting from the GIOP messages and how the performance metrics representing the system state are computed.

Event representation. Each GIOP message intercepted by CASPER becomes an event feeding the CEP engine of the performance metrics computation component. Each event contains (i)*Request ID*: The identifier of a request-reply interaction between two CORBA entities; (ii)*Message Type*: A field that characterizes the message and that can assume different values (e.g., Request, Reply, Close Connection) and (iii)*Reply Status*: It specifies whether there were exceptions during the request-reply interaction and, if so, the kind of the exception. In addition, we insert into the event further information related to the lower level protocols (TCP/UDP) such as source and destination IP, port, and timestamp. In order not to capture sensitive information of the ATC system (such as flight plans or routes), CASPER ignores the payload of the messages.

Performance metrics. Events sent to the CEP engine are correlated online so as to produce so-called performance metrics. After long time of observations of several metrics of the ATC CORBA-based system, we identified the following small set of metrics that well characterize the system, showing a steady behavior in case of absence of faults, and an unstable behavior in presence of faults:

- *Round Trip Time:* elapsed time between a request and the relative reply;
- *Rate of the messages carrying an exception:* the number of reply messages with exception over the number of caught messages;
- *Average message size:* the mean of the messages size in a given spatial or temporal window;
- *Percentage of Replies:* the number of replies over the number of requests in a given spatial or temporal window;
- *Number of Requests without Reply:* the number of requests expecting a reply that, in a given temporal window, do not receive the reply;
- *Messages Rate:* the number of messages exchanged in a given spatial or temporal window.

To compute the performance metrics we correlate the sniffed network data using the CEP engine ESPER [6]. This choice is motivated by its low cost of ownership compared to other similar systems (e.g. [9]) and its offered usability.

7.4. CASPER experimental evaluation

The first part of the evaluation on the field has been to collect a large amount of network traces from the ATC underlying communication network when in operation. These traces represented steady state performance behaviors. Additionally, on the testing environment of the ATC system we stressed some of the nodes till achieving software failure conditions, and we collected the relative traces. In our test field, we consider one of the nodes of the ATC system to be affected by either Memory or I/O stress (according to the experience of the ATC designers, these two stress conditions are typical of the observed system). After collecting all these traces, we trained CASPER. At end of the training phase, we deployed CASPER again on the testing environment of the ATC system in order to conduct experiments in operative conditions. Our evaluation assesses the system state inference component accuracy and the failure prediction module accuracy (see Figure 7). In particular, we evaluate the former in terms of N_{tp} (number of true positives) the system state is unsafe and the inferred state is "system unsafe"; N_{tn} (number of true negatives): the system state is safe and the inferred state is "system safe"; N_{fp} (number of false positive): the system state is safe but the inferred state is "system unsafe"; and N_{fn} (number of false negatives): the system state is unsafe but the inferred state is "system safe". Using these parameters, we compute the following metrics that define the accuracy of CASPER:

- Precision: $p = \frac{N_{tp}}{N_{tp}+N_{fp}}$

- Recall (or true positive rate): $r = \frac{N_{tp}}{N_{tp}+N_{fn}}$

- F-measure: $F = 2 \times \frac{p \times r}{p+r}$

- False Positive Rate: $f.p.r. = \frac{N_{fp}}{N_{fp}+N_{tn}}$

We evaluate the latter module in terms of N_{fp} (number of false positive): the module predicts a failure that is not actually coming and N_{fn} (number of false negatives): the module does not predict a failure that is coming. **Testbed.** We deployed CASPER in a dedicated host located in the same LAN as the ATC system to be monitored (see Figure 7). This environment is actually the testing environment of the ATC system where new solutions are tested before getting into the operational ATC system. The testing environment is composed by 8 machines, 16 cores 2.5 GHz CPU, 16 GB of RAM each one. It is important to remark that CASPER does not know the application nor the service logic nor the testbed details.

7.5. Faults and failures

The ATC testbed includes two critical servers: one of the server is responsible for disk operations (I/O) and another server is the manager of all the services. In order to induce software failures in the ATC system, we apply the following actions in such critical servers: (i)*memory stress*; that is, we start a memory-bound component co-located with the manager of all ATC services, to grab constantly increasing amount of memory resource; (ii)*I/O stress*; that is, we start an I/O-bound component co-located with the server responsible for disk operations, to grab disk resources. In both cases we brought the system to the failure of critical services. During the experiment campaign, we also considered the CPU stress; however, we discovered that due to the high computational power of the ATC nodes, the CPU stress never causes failures.

7.6. Results of CASPER

We run two types of experiments once CASPER was trained and tuned. In the first type, we injected the faults described in previous section in the ATC testing environment and we carried out 10 tests for each type of fault. In general, we obtained that in the 10 tests we carried out, the time-to-failure in case of memory stress varied in the range of [183s, 216s] and the time-to-prediction in the range of [20.8s, 27s]. In case of I/O stress, in the 10 tests, the time-to-failure varied in the rage of [353s, 402s] whereas the time-to-prediction in the range of [19.2s, 24.9s]. The time-to-failure in our evaluation has been long enough in order to trigger proper countermeasures, that can be set before the failure, to either mitigate damages or enable recovery actions. Further details can be found in [1].

8. Conclusion

This chapter presented the motivations that led the current literature to develop novel solutions to failure management in ATC systems. We analyzed the failure management objectives, what the faults and failures are and how they can be managed in a real ATC system. Some hint on the failure management reactive and proactive approaches have been described and two case studies have been presented: a reactive approach, that uses FT-CORBA, today in operation and a novel proactive approach that uses a combination of Complex Event Processing and Hidden Markov Models to predict the occurrence of failures in ATC systems.

Author details

Luca Montanari and Roberto Baldoni
"Sapienza" University of Rome, Italy

Fabrizio Morciano and Marco Rizzuto
"Selex Sistemi Integrati" a Finmeccanica Company, Italy

Francesca Matarese
"SESM" a Finmeccanica Company, Italy

9. References

[1] Baldoni, R., Lodi, G., Montanari, L., Mariotta, G. & Rizzuto, M. [2012]. Online black-box failure prediction for mission critical distributed systems, *to appear in proceedings of SAFECOMP 2012*, Springer Berlin / Heidelberg.

[2] CARDAMOM [website]. Cardamom middleware website. http://www.cardamom.eu/.

[3] EC482 [2008]. Commission regulation (ec) no 482/2008, *Official Journal of the European Union* pp. 5–9.

[4] Elnozahy, E. N., Alvisi, L., Wang, Y.-M. & Johnson, D. B. [2002]. A survey of rollback-recovery protocols in message-passing systems, *ACM Comput. Surv.* 34(3): 375–408.

[5] ESARR6. [2010]. *ESARR 6. EUROCONTROL Safety Regulatory Requirement. Software in ATM Systems.*, 2.0 edn, European Organisation for the Safety of Air Navigation.

[6] Esper [2012]. Esper project web page. http://esper.codehaus.org/.

[7] Fischer, M. J., Lynch, N. A. & Paterson, M. [1985]. Impossibility of distributed consensus with one faulty process, *J. ACM* 32(2): 374–382.

[8] Gertler, J. [1988]. Survey of model-based failure detection and isolation in complex plants, *Control Systems Magazine, IEEE* 8(6): 3 –11.

[9] IBM [2011].   System S Web Site.   `http://domino.research.ibm.com/comm/`
`research_projects.nsf/pages/esps.index.html`.

[10] Kapadia, R., Stanley, G. & Walker, M. [2007]. Real world model-based fault management.,
18th International Workshop on the Principles of Diagnosis Nashville TN.

[11] Laprie, J.-C., Avizienis, A., Randell, B. & Landwehr, C. E. [2004].   Basic concepts and
taxonomy of dependable and secure computing, *IEEE Trans. Dependable Sec. Comput.*
1(1): 11–33.

[12] Liebert [2005]. *Regulatory Compliance and Critical System Protection*, Liebert Corporation.

[13] Luckham, D. C. [2001]. *The Power of Events: An Introduction to Complex Event Processing
in Distributed Enterprise Systems*, Addison-Wesley Longman Publishing Co., Inc., Boston,
MA, USA.

[14] OMG [CCM]. Corba component model (ccm), omg specification, formal/2011-11-03, part
3 - components. `http://www.omg.org/spec/CORBA/3.2/Components/PDF`.

[15] OMG [FT-CORBA]. Fault tolerant corba (ft), omg specification, formal/2010-05-07 , v1.0.
`http://www.omg.org/spec/FT/1.0/PDF`.

[16] OMG [LTLOAD].   Lightweight load balancing service (ltload), omg specification,
formal/2010-02-04, v1.0. `http://www.omg.org/spec/LtLOAD/1.0/PDF`.

[17] OMG [LWFT].   Lightweight fault tolerance for distributed rt systems (lwft),
ptc/2011-06-05, beta 2. `http://www.omg.org/spec/LWFT/1.0/Beta2/PDF`.

[18] OMG [website]. Object management group webpage. `http://www.omg.org/`.

[19] Rabiner, L. & Juang, B. [1986].   An introduction to hidden markov models, *ASSP
Magazine, IEEE* 3(1): 4 – 16.

[20] Salfner, F. [2008].   *Event-based Failure Prediction: An Extended Hidden Markov Model
Approach*, PhD thesis, Department of Computer Science, Humboldt-Universität zu Berlin,
Germany.

[21] Salfner, F., Lenk, M. & Malek, M. [2010]. A survey of online failure prediction methods,
ACM Computing Surveys (CSUR) 42(3): 1–42.

[22] Simpson, W. & Sheppard, J. [1994]. *System test and diagnosis*, Kluwer Academic.
URL: *http://books.google.it/books?id=Pjr93wWJMiQC*

[23] Stanley, G. M. & Vaidhyanathan, R. [1998].   A generic fault propagation modeling
approach to on-line diagnosis and event correlation., *3rd IFAC Workshop on On-line Fault
Detection and Supervision in the Chemical Process Industries*,.

[24] Trivedi, K. S. & Vaidyanathan, K. [2008].   Software aging and rejuvenation, *Wiley
Encyclopedia of Computer Science and Engineering*.

[25] Williams, A. W., Pertet, S. M. & Narasimhan, P. [2007].   Tiresias: Black-box failure
prediction in distributed systems, *Proc. of IEEE IPDPS 2007*, Los Alamitos, CA, USA.

Chapter 8

The Autonomous Flight

Tone Magister and Franc Željko Županič

Additional information is available at the end of the chapter

1. Introduction

The Autonomous Flight Airspace (AFA) [9] is the evolutional offspring of the Free Flight Airspace (FFA), and enabler of integrated flight operations of aircraft with autonomous flight capabilities (for instance, Unmanned Aircraft Systems (UAS)).

In the FFA the responsibilities for the airborne spacing and separation assurance are delegated to flight crews on board the aircraft, and the ground–based Air Traffic Management (ATM) is to resume separation authority in emergencies only [2]. Therefore, humans are the decision–makers, as well as operatives in the FFA.

Since airborne separation assurance is a fundamental principle of the FFA and the Airborne Separation Assurance System (ASAS) its main enabler, the AFA introduces the autonomous airborne separation assurance with Autonomous–ASAS (AASAS). The AFA is marked by the machine–based decision–making, and the AFA is restricted to the ASAS and AASAS equipped aircraft but both types with autonomous flight capabilities. In the future the only humans–in–the–loop conducting flight operations through AFA are going to be ground–based UAS operators, air traffic flow managers of the next generation ATM, and systems supervisors (*pilots of present–day terminology*) onboard remnant "old–school" manned aircraft. Based on 4D trajectory planning the AASAS concept covers machine–based (a) traffic situational awareness, and (b) airborne spacing and self–separation assurance through (c) autonomous in–flight conflict detection and resolution.

The AFA concept is not only important for implementation of non–segregated UAS flight operations [8], but also for the future air transport system responding to the society's emerging needs (which are not limited to enabling permeability of increasing volume of air traffic, but include other issues; i.e., for example airborne security when it is necessary for the pilot to transfer his/her responsibilities to an automatic system due to a hijack situation for flight trajectory protection and safe automatic return to the ground as envisioned in [1]).

Analogous to the traffic complexity at both highway ends, air traffic is inherently complex especially in both zones adjacent to the boundary between the AFA (or FFA) and non–autonomous (or un–free) flight airspace (non–AFA). The number (quantum) of conflicts between aircraft is proportional to the complexity of the in–flight traffic situation [7]. For AFA implementation (or any airspace organization with changing delegation of responsibilities for the airborne spacing and separation assurance), the transition flights between the AFA and non–AFA therefore represent a critical safety issue.

The prediction accuracy of the future trajectory of each and every aircraft aloft drive the stability of the predicted four-dimensional traffic situation in the airspace confined with a look-ahead time and consequently the ASAS and/or AASAS on-board each aircraft ability to detect every potential conflicton-time and correctly. The statement holds for either free-flight [2], sector-less airspace [4], or automated airspace [5] envisioned to cope with the increasing demand in the crowded skies above.

With the existing technology and methodology the look-ahead time for the construction of accurate future aircraft flight trajectory is reduced to only about 5 to 7 minutes in advance. Prolonged look-ahead timeusing the current airborne separation assurance technology, designed not to include the future intent,results in predicted traffic situation instability [6] and consequently in conflict detection unreliability or even inability.

The study is also focused on the design of an advanced four-dimensionalmodel of aircraft relative flightproviding the capabilities of the AASAS to detect conflicts beyond the borders of the AFA and enabling Autonomous Flight concept implementation.

2. Autonomous flight airspace

2.1. Problem of transition flight to/from autonomous flight airspace

The complexity of air traffic and quantum of in–flight conflicts between aircraft in the AFA (or FFA) and its non–AFA neighborhood can be investigated using the theory of airspace fractal dimensions proposed by Mondoloni and Liang in [11]. This theory was introduced as methodology capable of simultaneously distinguishing between complexity of air traffic situation as a consequence of management of air traffic flow, and complexity of an air traffic situation as a consequence of geometry and organization of airspace. Fractal dimension is a number $D \in \{D \in \mathbb{R}, 1 \leq D \leq 3\}$ assigned to the particular flight corresponding to freedom of aircraft motion. As shown in Table 1, with increasing freedom of movement the fractal dimension of flight increases, and vice versa.

The frequency of in–flight conflicts decreases exponentially if fractal dimension of aircraft flight increases. Alternatively, the number of in–flight conflict encounters C threatening aircraft ($dC = CR\,dt$) increases with decreasing freedom of its flight ΔD, and their relation can be approximated from data of Table 1 as:

$$\int CR\,dt \approx 11.472 - 2.452\Delta D \tag{1}$$

Scenario / Degrees of Freedom				Fractal Dimension
Airway Network	Heading	Airspeed	Top of Descent	
D	D	D	D	1.0
D	n/a	F	n/a	2.0
e	F	D	n/a	2.0
e	F	n/a	D	2.0
n/a	D	F	F	2.1
e	F	n/a	F	2.4
e	F	F	n/a	2.6
Controlled Airspace / Transition Areas *form 10,000ft to FL410; 120nm box around airport*				1.22 – 1.39
Upper Control Area *upper airspace above FL240*				1.13 – 1.32
TMA/CTA – Terminal and Control Area *from 10,000ft to FL240*				1.2

D	Determined parameter
F	Free parameter
e	Exclusive parameter
n/a	Undefined parameter

Table 1. The fractal dimension of flight is proportional to the freedom of movement (data compiled from [11]).

Since descending and/or climbing aircraft through the sector of level cruising flights markedly increase the air traffic controller's workload [3], and consequently decrease sector throughput, earlier studies such as [2] anticipated mostly level transition flights from the FFA (and applicable to the AFA as well). Level transition flights to and from the AFA (or FFA) require that the AFA (or FFA) and the controlled airspace (CA) are positioned side by side. Such airspace configuration again increases the air traffic controller's workload by introducing the mix of differently equipped aircraft subjected to essentially different procedures, namely the mix of controlled flights and autonomous (or free) flights en–route to or from the AFA (or FFA). However, to simultaneously gain increased airspace capacity and flight economics together with decreased emissions from optimized flights, the AFA (or FFA) should and will extend above CA (Fig. 1). Obviously there is more than one reason to consider transition to and from the AFA (or FFA) while aircraft are climbing or descending.

The transition flight from the AFA into the CA results in significantly decreased freedom of flight; aircraft path might be dictated directly by ATM or at least to a certain extent confined by the route network. Due to decrease in freedom of flight the fractal dimension of aircraft path will decrease (Table 1) and consequently the in–flight conflict encounter for transitioning aircraft will inevitability increase (1). The greater the differences between fractal dimensions of flight in the AFA and CA are, the greater is the increase in conflict encounter for transitioning aircraft at the boundary between the AFA and CA.

The greatest (50%) decrease of fractal dimension of flight and resulting drastic 135% increase of conflict encounters (1) occurs, when an aircraft transits the border between the AFA (or

FFA) and the CA through the arbitrary place (TC) at level flight and enters directly into the network of airways of CA.

The solution to the problem of transition flight conflict encounter increase at the border between the AFA (or FFA) and the CA consists of:

1. the transition to and from the AFA (or FFA) into the CA in non–level flight; i.e., introduction of transition in descent and climb;
2. gradually decreased degree of freedom of flight in the direction from the AFA (or FFA) into the CA; i.e., progressively dictated parameters of flight along the transitioning route before an aircraft leaves the AFA (or FFA), upon leaving the AFA, and afterwards while flying in the CA (and vice versa for the flight in the opposite direction transitioning to the AFA (or FFA));
3. CA organization and air traffic flow regulation adequate to the procedures of autonomously (or free) flying aircraft entering from the AFA (or FFA) and mixing with the rest of the traffic.

The proposed solution to the transitioning flight problem is presented in Fig. 1. Drastic increase in conflict encounter imminent to an aircraft at the AFA border in level transition flight is dispersed. In consequence, severity of conflict encounters along its transitioning trajectory is reduced.

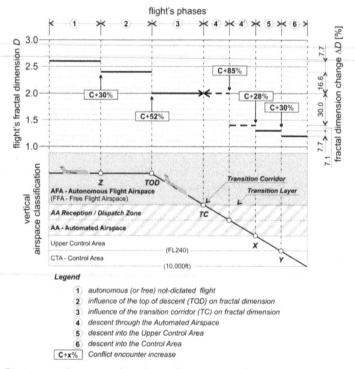

Figure 1. Conflict increase dispersion along descending transition from the AFA.

The top of descent (TOD) determination in the AFA has a two–fold impact on the freedom of flight decrease while an aircraft is still flying in the AFA. Determination of the TOD in the AFA itself (Table 1), as the trajectory determination factor, reduces the fractal dimension of flight even before an aircraft reaches the edge of the AFA (Z; Fig. 1). Furthermore the TOD can only be determined by the intersection of an aircraft cruising level and the trajectory of its descent (with a constant rate of descent to the assigned destination) through the rest of the transitioning aircraft free transition corridor (TC; Fig. 1) closest to the optimal route through the border between the AFA and CA. The transition corridor (TC) is a one–way passage in the transition layer through which an aircraft flies from the AFA into the CA (and vice versa); at the same time it is the starting point of a particular airway in the CA. The TC defines the four–dimensional position of an aircraft transitioning from the AFA and direction of flight in the CA adjacent to the transition layer; consequently the TC is the restrictive factor which decreases freedom of aircraft movement and the fractal dimension of its flight (Table 1). Recurrently determined TOD and TC define the route of an aircraft leaving the AFA, and by gradually decreasing its freedom of flight dispersing threatening conflict encounters with neighboring aircraft along its way.

Descending from the AFA via the TC an aircraft enters the CA. In the part of the CA that borders upon the AFA it is of a critical importance that the autonomous flights can be safely integrated with the rest of not–autonomous traffic, and that the airspace organization including traffic flow management enables a fractal dimension comparable to the fractal dimension of the AFA. Both major criteria of the CA bordering the AFA are met with the Automated Airspace (AA) type of CA proposed in [4], if a reception zone is introduced into the AA at its border with the AFA.

The AA is based upon the ground–based automation system that provides in–flight separation assurance via data–link communication for properly equipped aircraft. The ground–based automation system issue clearances for aircraft intended trajectories and/or it can up–load safe trajectories directly into the flight management system module of the AASAS of the autonomous aircraft and ASAS of the non–autonomous aircraft.

The roles of controllers in the AA are strategic control of traffic flow, handling of unusual traffic situations, and monitoring and control of unequipped aircraft [4]. This facilitates the autonomous and non–autonomous aircraft mix in the AA.

The reception zone is an integral part of the AA adjacent to the border with the AFA (flying in the opposite direction an aircraft leaves the AA through the dispatch zone). The concourse pattern of airways starting in the AA reception zone with each TC at the border with the AFA is adjusted to match the directions of cardinal routes of the AFA; their geometry and organization serves as a collector of air traffic flows from various TCs onto the main central airways of the AA. The airways structure and management of traffic flow-- i.e., the aircraft trajectory control and restrictions--are such that the fractal dimension of the AA reception zone corresponds to the AFA fractal dimension. The fractal dimension of a flight upon crossing the border between the AFA and AA reception zone remains unchanged allowing that in the most critical part of the transition flight in the vicinity of the

TC and in the TC the conflict encounter does not increase for the transitioning aircraft (Fig. 1).

Descending through the CA below the AA the aircraft traverse airspace sectors of different classes with progressively increased restrictions and control (dictation) of its trajectory each time the sector boundary is crossed, leading to non–severe but gradual increase in conflicts in succession of each sector boundary crossing (X,Y; Fig. 1). However the greatest fractal dimension of a non–AA CA is far less than the AA reception zone fractal dimension. Consequently the greatest (30%) change of fractal dimension of a transitioning flight is expected to occur in the AA resulting in an 85% increase in conflict encounter (1) threatening descending aircraft (Fig. 1).

The challenge of the AA organization and traffic flow regulation is to progressively dictate the flight of the transitioning aircraft to secure gradual decrease in AA fractal dimension in the direction away from the transition layer from the value corresponding to the AFA fractal dimension with a value similar to the upper CA fractal dimension. That way the expected increase in conflict is dispersed further along the entire descending trajectory through the AA. The spacing and separation assurance actors in the AA are the AASAS of the autonomous aircraft, the ASAS of a free–flying aircraft, crews of unequipped aircraft, the ground–based separation assurance automation system, and the AA strategic traffic flow controller; but parallel to the human error hazard, a data–link communication failure imposes the greatest risk for flight safety in the AA.

2.2. Autonomous flight airspace design

For the safety of aircraft flying in the AFA and AA, both are demarcated by the transition layer (TL), defined by the entry and exit plane that are separated at least by the vertical separation minimum. The AFA extends above the entry plane, while the AA is positioned below the exit plane (Fig. 2). Aircraft are transitioning to and from the AFA through the bordered tube–like transition corridor (TC) at the TL.

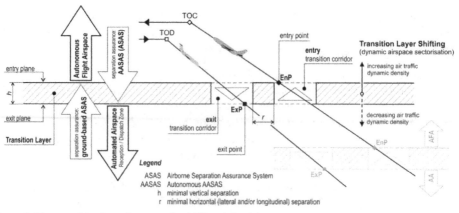

Figure 2. The transition layer between the AFA and the AA.

Aircraft flows from either side of the TL converging for transit through the TCs, leading to the traffic dynamic density increase on either side of the TL in its proximity (applying the WJHTC/Titan Systems Metric: the convergence recognition index, separation critically index, and degrees of freedom index will be the most critical [7]). Traditionally the traffic dynamic density is limited by the air traffic controller workload, however even in the AFA or AA the dynamic density will still remain limiting factor due to the limited airborne and ground–based separation assurance system processor capacity as well as limited data–link bandwidth. Dynamic airspace sectorization will ensure that the air traffic dynamic density doesn't reach its limits by the TL shifting. The (pressure) altitude of the TL is proportional to the air traffic dynamic density trend; if it is increasing, for example, the TL will ascend, resulting in AA vertical expansion simultaneously with the AFA *contraction* (Fig. 2).

In the AFA and AA aircraft, in–flight spacing and separation relies on the machine–based decision–making ASAS; in the AFA the AASAS, responsibility extends to the exit plane of the TL, while in the AA the ground–based automation ASAS responsibilities extend to the entry plane of the TL. Since the exit plane doesn't coincide with the entry plane the airborne spacing and separation of aircraft responsibilities are shared in the TL between the AASAS onboard autonomous aircraft and the ground–based automation ASAS of the AA. Due to shared responsibilities for airborne separation the entrance and exit TCs must be separated; aircraft are flying from the AFA through the exit TC, while they are entering the AFA through the entry TC (Fig. 2). Consequently, and considering again the fact that airborne separation is based upon the machine–based decision–making in the AFA and AA, any conflict avoidance maneuvering can only be coordinated implicitly between the AASAS and/or ASAS onboard aircraft involved in the conflict encounter, including implicit coordination of future 4D trajectories of aircraft in the area.

Since AASAS of autonomously flying aircraft is still responsible for the in–flight spacing and separation when the transitioning aircraft is in the TC at the exit plane of the TL, the AASAS has to be capable of detecting possible conflict situations with aircraft flying in the AA even before the time of transition from the AFA. Actually the rest of the transitioning aircraft-free--and especially conflict-free--exit TC can only be selected (determined) in the process of aircraft descent trajectory from the AFA definition before the TOD is reached, providing that accurate prediction of along the descending route and across the TL airborne traffic situation can be made. The greater look–ahead time for accurate and stable 4D prediction of an airborne traffic situation, demands an accurate model of aircraft future relative positions based on their real future ground speeds, their future intents, as well as on future area weather conditions. (Similar requirements are applicable for the ground–based ASAS of the AA since it is responsible for airborne aircraft spacing and separation in the TC at the entry plane of TL.)

Rules–of–the–sky tailored to the AFA flight operations are necessary for competitive rivalry for the best optimal trajectory prevention, and also due to the fact that conflict avoidance maneuvering can only be coordinated implicitly between AASAS and/or ASAS onboard aircraft. For the transition flight to and from the AFA safety, a pair of rules applies. *The priority flight (first) rule*: "An aircraft that flies lower than the other aircraft involved in the conflict encounter when conflict is detected, has the right–of–way." The rule therefore implies that only the higher flying aircraft is responsible for resolving the conflict situation. Since the AFA extends above the AA, the autonomous aircraft flying in the AFA are obliged to maneuver for menacing conflict resolution in case they are encountering conflict with an aircraft climbing to enter the AFA from the AA, and in their envisioned descent transition from the AFA. This priority rule also defines minimum separation between the entry and exit plane, as well as minimum separation between the entry and exit TCs at the TL for unnecessary aircraft maneuvering in the AFA prevention. A *maneuver flight (second) rule*: "After a conflict is detected, it is prohibited for the aircraft which has the right–of–way to alter planned trajectory until conflict is resolved." A pair of rules is therefore defined to ensure reliable implicit coordination of conflict avoidance maneuvering and increase conflict resolution predictability.

3. Autonomous airborne separation assurance

3.1. Aircraft relative flight model

The primitive flight model predicts each aircraft's future trajectory with extrapolation of its ground speed vector from aircraft's last position, while the aircraft's ground speed vector is derived with interpolation between its last two known positions. The predictions of this model are therefore based upon a set of presumptions of: (a) constant aircraft ground speed and direction, followed by (b) constant wind speed and direction, and (c) constant static state (temperature) of atmosphere as well.

Let's investigate the reliability of conflict detection in the encounter situation between our own aircraft denoted by index 2 in descent and the intruder denoted by index 1 in level flight below. If the time of the present level cruise phase of own flight is denoted as τ, while t denotes the time of own aircraft descent as a subsequent phase of its flight, then the primitive model of relative position between own aircraft and intruder with the top of descent accounted for as a future intent can be written as:

$$\dot{x}_R(\tau,t) = v_{2G} \cos\psi_R - v_{1G}$$
$$\dot{y}_R(\tau,t) = v_{2G} \sin\psi_R \qquad\qquad (2)$$
$$\dot{z}_R(\tau,t) = -v_{2G} \tan\theta_R(\tau,t)$$

where ψ_R and θ_R are the relative angles between aircraft trajectories in the horizontal and vertical plane successively, and v_G is their ground speed.

Closer examination of a primitive model (2) reveals that, if there is no distinction made between the time of own aircraft level flight and the time of its descent then, the primitive model cannot predict when this aircraft will alter its trajectory. Furthermore, following presumptions of the primitive model described above it is obvious that, even if deficiency of this primitive model is corrected with introduction of each aircraft future intent, this primitive model cannot account for the future aircraft trajectory variations due to the true airspeed variableness as a function of a non-zero vertical static temperature gradient in the troposphere.

The improved model of aircraft flight was derived to include not only the aircraft (crew) future intent but also to consider:

a. the aircraft true airspeed v variableness $v = v(\vartheta_S(z))$ as a function of static state of an atmosphere ϑ_S (i.e. static air temperature (SAT) T; $\vartheta_S(z) = \{T(z)\}$) variation with aircraft height z above reference,

b. the aircraft true airspeed variableness $v = v(\sigma(z))$ due to the changing set of speed regimes $\sigma(z) = \{M, v_C\}$ of descent and/or climb with constant Mach number M and/or with constant calibrated airspeed v_C,

c. the influence of the dynamic state of an atmosphere $\vartheta_D(z) = \{w(z), \xi(z)\}$ defined with the wind speed w and direction ξ on the progressive speed V of an aircraft which can be written as $V = V(v, \vartheta_D(z))$.

Based on the simplification that an angular velocity vector of each aircraft equals to zero, and an assumption that an alteration of each aircraft trajectory is instantaneous (chapter 3.2.1), the improved model of aircraft relative motion is defined with:

$$
\begin{aligned}
x'_R &= V_2(v_2(\sigma_2(z), \vartheta_S(z)), \vartheta_D(z))\cos\theta_R(\sigma_2(z))\cos\psi_R - \\
&\quad - V_1(v_1(\sigma_1(z), \vartheta_S(z)), \vartheta_D(z)) \\
y'_R &= V_2(v_2(\sigma_2(z), \vartheta_S(z)), \vartheta_D(z))\sin\theta_R(\sigma_2(z))\cos\psi_R \\
z'_R &= V_2(v_2(\sigma_2(z), \vartheta_S(z)), \vartheta_D(z))\sin\theta_R(\sigma_2(z))
\end{aligned}
\tag{3}
$$

andcan be transformed into the time dependant function using the rate of climb (+) or descent (−) definition:

$$
\pm\frac{dz}{dt} = V(z)\sin\theta_R(\sigma_2(z))
\tag{4}
$$

where the progressive speed V of an aircraft follows form the aircraft speed vector triangle:

$$
V(z) = \frac{w\cos(\xi - \psi) + \sqrt{v^2(\sigma(z), \vartheta_S(z)) - w^2\sin^2(\xi - \psi)}}{\cos\theta_R(\sigma_2(z))}
\tag{5}
$$

The solution of the improved kinematical model of aircraft relative flight (3; (4) and (5)) is presented, as improved model of the aircraft relative motion, for the case that aircraft abbreviated as A2 and its flight parameters denoted by index 2, starts its descent at the top of descent (TOD) from a cruise level z_{FL2} in stratosphere $z_{FL2} > z_{tp}$ (tropopause at z_{tp}) at $t_{TOD} > t_0$ after conflict is detected at t_0, while the intruder denoted by index 1 continues its constant Mach number M level cruise. The solution of (3) provided is partitioned according to descending aircraft flight phases; note that s, c and t denote trigonometric functions of sine, cosine and tangent.

a. $t_0 \leq t \leq t_{TOD}$ ($t_0 = 0$): the A2 is in a $M_2 = const$ level cruise ($\theta_2 = 0$) in the stratosphere:

$$x_R(t) = x_R(t_0) + \left(w \left(c_{\lambda_2} c_{\psi_R} - c_{\lambda_1} \right) - \sqrt{M_1^2 a_{FL1}^2 - w^2 s_{\lambda_1}^2} + c_{\psi_R} \sqrt{M_2^2 a_{tp}^2 - w^2 s_{\lambda_2}^2} \right) t$$

$$y_R(t) = y_R(t_0) + s_{\psi_R} \left(w c_{\lambda_2} + \sqrt{M_2^2 a_{tp}^2 - w^2 s_{\lambda_2}^2} \right) t \tag{6}$$

$$z_R(t) = z_R(t_0)$$

where $r_R(t_0) = (x_R(t_0), y_R(t_0), z_R(t_0))$ is the initial aircraft relative position when conflict is detected at t_0.

b. $t_{TOD} < t \le t_{tp}$: the A2 descends in constant M speed–regime with constant angle of descent $\theta_R = \theta_2$ through the stratosphere:

$$x_R(t) = x_R(t_{TOD}) + \left(w \left(c_{\lambda_2} c_{\psi_R} - c_{\lambda_1} \right) - \sqrt{M_1^2 a_{FL1}^2 - w^2 s_{\lambda_1}^2} + c_{\psi_R} \sqrt{M_2^2 a_{tp}^2 - w^2 s_{\lambda_2}^2} \right) t$$

$$y_R(t) = y_R(t_{TOD}) + s_{\psi_R} \left(w c_{\lambda_2} + \sqrt{M_2^2 a_{tp}^2 - w^2 s_{\lambda_2}^2} \right) t \tag{7}$$

$$z_R(t) = z_R(t_{TOD}) - t_{\theta_2} \left(w c_{\lambda_2} + \sqrt{M_2^2 a_{tp}^2 - w^2 s_{\lambda_2}^2} \right) t$$

where $r_R(t_{TOD})$ is solution of (6) for $t = t_{TOD}$.

c. $t_{tp} < t \le t_p$: after passing the tropopause at t_{tp} the A2 descends in constant M speed–regime through the troposphere:

$$x_R(t) = x_R(t_{tp}) + \left[w \left(c_{\lambda_2} c_{\psi_R} - c_{\lambda_1} \right) + c_{\psi_R} k_5 M_2 \sqrt{\chi R} - \sqrt{M_1^2 a_{FL1}^2 - w^2 s_{\lambda_1}^2} \right] (t - t_{tp}) + c_{\psi_R} k_6 M_2 \sqrt{\chi R} \left(t^2 - t_{tp}^2 \right)$$

$$y_R(t) = y_R(t_{tp}) + s_{\psi_R} (t - t_{tp}) \left(w c_{\lambda_2} + k_5 M_2 \sqrt{\chi R} \right) + s_{\psi_R} k_6 M_2 \sqrt{\chi R} \left(t^2 - t_{tp}^2 \right) \tag{8}$$

$$z_R(t) = z_R(t_{tp}) - t_{\theta_2} (t - t_{tp}) \left(w c_{\lambda_2} + k_5 M_2 \sqrt{\chi R} \right) - t_{\theta_2} k_6 M_2 \sqrt{\chi R} \left(t^2 - t_{tp}^2 \right),$$

where $r_R(t_{tp})$ is solution of (7) for $t = t_{tp}$, while k_5, k_6, and k_1 and k_2 are:

$$k_5 = \sqrt{T_{OR} + \frac{L}{2k_2} \left[k_1 - \sqrt{k_1^2 + 4k_2 \left(t_{tp} t_{\theta_2} + z_{tp} \left(k_1 + k_2 z_{tp} \right) \right)} \right] - \frac{w^2 s_{\lambda_2}^2}{M_2^2 \chi R}} \tag{9}$$

$$k_6 = L t_{\theta_2} \left(4 k_5 \sqrt{k_1^2 + 4k_2 \left(t_{tp} t_{\theta_2} + z_{tp} \left(k_1 + k_2 z_{tp} \right) \right)} \right)^{-1} \tag{10}$$

$$k_1 = \left(w c_{\lambda} + \sqrt{M^2 \chi R T_{OR} - w^2 s_{\lambda}^2} \right)^{-1} \tag{11}$$

$$k_2 = \frac{k_1^2 M L \sqrt{\chi R}}{4} \left(T_{OR} - \frac{w^2 s_{\lambda}^2}{M^2 \chi R} \right)^{-\frac{1}{2}} \tag{12}$$

d. $t > t_p$: at t_p the A2 changes its speed–regime and continues its descent through the troposphere with the constant calibrated airspeed ($v_{C2} = const$):

$$x_R(t) = x_R(t_p) + \left[w\left(c_{\lambda_2} c_{\psi_R} - c_{\lambda_1} \right) + c_{\psi_R} k_7 - \sqrt{M_1^2 a_{FL1}^2 - w^2 s_{\lambda_1}^2} \right](t - t_p) + c_{\psi_R} k_8\left(t^2 - t_p^2\right)$$

$$y_R(t) = y_R(t_p) + s_{\psi_R}\left(t - t_p\right)\left(w c_{\lambda_2} + k_7\right) + s_{\psi_R} k_8\left(t^2 - t_p^2\right) \tag{13}$$

$$z_R(t) = z_R(t_p) - t_{\theta_2}\left(t - t_p\right)\left(w c_{\lambda_2} + k_7\right) - t_{\theta_2} k_8\left(t^2 - t_p^2\right),$$

where $r_R(t_p)$ is solution of (8) for $t = t_p$, while k_7, k_8, k_9, k_{10}, and k_3, k_4 are:

$$k_7 = \sqrt{\frac{2\chi R T_{0R}}{\chi - 1}\left(1 + \frac{Lk_9}{2k_4 T_{0R}}\right)\left[\left(k_{10}\left(1 + \frac{Lk_9}{2k_4 T_{0R}}\right)^{-\frac{g_0}{LR}} + 1\right)^{\frac{\chi - 1}{\chi}} - 1\right] - w^2 s_{\lambda_2}^2} \tag{14}$$

$$k_8 = \frac{\chi R L t_{\theta_2}}{2(\chi - 1)(k_3 - k_9)k_7}\left[-1 + \left(k_{10}\left(1 + \frac{Lk_9}{2k_4 T_{0R}}\right)^{-\frac{g_0}{LR}} + 1\right)^{\frac{\chi - 1}{\chi}}\left(1 - \frac{(\chi - 1)g_0 k_{10}}{\chi R L}\left(k_{10} + \left(1 + \frac{Lk_9}{2k_4 T_{0R}}\right)^{\frac{g_0}{LR}}\right)^{-1}\right)\right] \tag{15}$$

$$k_9 = k_3 - \sqrt{k_3^2 + 4k_4\left(t_{\theta_2} t_p + z_p\left(k_3 + k_4 z_p\right)\right)} \tag{16}$$

$$k_{10} = \left(1 + \frac{\chi - 1}{2}\left(\frac{v_{C2}}{a_0}\right)^2\right)^{\frac{\chi}{\chi - 1}} - 1 \tag{17}$$

$$k_3 = \left(w c_\lambda + \sqrt{\frac{v_C^2 T_{0R}}{T_0} - w^2 s_\lambda^2}\right)^{-1} \tag{18}$$

$$k_4 = \frac{k_3^2}{4}\left(\frac{v_C^2 T_{0R}}{T_0} - w^2 s_\lambda^2\right)^{-\frac{1}{2}}\left(\frac{v_C^2 L}{T_0} - 2g_0\left(1 + \frac{\chi - 1}{2}\left(\frac{v_C}{a_0}\right)^2\right)\left[1 - \left(1 + \frac{\chi - 1}{2}\left(\frac{v_C}{a_0}\right)^2\right)^{-\frac{\chi}{\chi - 1}}\right]\right) \tag{19}$$

The abbreviations not given in the text are: g_0 is acceleration of gravity, L is (temperature atmospheric) lapse rate, R is universal gas constant, χ is ratio of specific heats, a_0 and a_{FL} are speed of sound at reference level of standard atmosphere and at aircraft flight level (FL), T_0 and T_{0R} are reference SATs of standard and real atmosphere, λ represents the difference between the wind direction ξ and aircraft true heading ψ, while index R denotes the relative parameter.

3.2. Accuracy of modeling

3.2.1. Simplifications based errors

At the top of descent (TOD) an aircraft starts its rotation $\Omega(t)=(0,\omega_\theta,0)$ (for $t\in[0,t_t]$) about its lateral axis until the angle of descent θ is established after transition time t_t as is shown in Fig. 3.

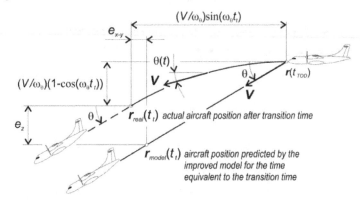

Figure 3. Simplified transition into descent.

For simplicity of an improved model of aircraft relative flight (3) the instantaneous aircraft transition into descent is assumed:

$$\lim_{t_t\to 0}\frac{V}{\omega_\theta}\left(1-\cos\left(\omega_\theta t_t\right)\right)=0$$

$$\lim_{t_t\to 0}\frac{V}{\omega_\theta}\sin\left(\omega_\theta t_t\right)=0$$

(20)

Due to simplification (20), the aircraft trajectory is not smooth at the TOD, resulting in the horizontal e_{x-y} and vertical e_z plane error of aircraft position prediction in the period of transition time $t\in[0,t_t]$. It can be theoretically estimated from Fig. 1 as:

$$e_{x-y}(t_t)=V(t_t)\left(\cos\theta-\frac{\sin\theta}{\theta}\right)t_t$$

$$e_z(t_t)=V(t_t)\left(\sin\theta-\frac{1-\cos\theta}{\theta}\right)t_t$$

(21)

The position errors e_{x-y} and e_z (21) are proportional to the transition time t_t, angle of descent θ, and aircraft progressive speed $V=f(v(\sigma(z),\vartheta_S(z)),\vartheta_D)$. They reach their maximum after the transition into descent is completed at t_t; however, after transition time t_t, the theoretical position errors e_{x-y} and e_z (21) of improved model (3) are constant. The theoretical position errors e_{x-y} and e_z are presented in Fig. 4 for constant Mach number speed regime transition into the descent with standard constant angle of $\theta=3°$.

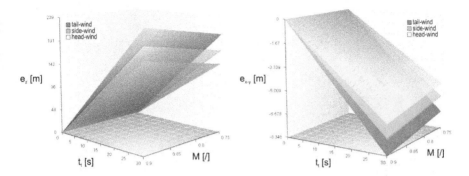

Figure 4. The improved model aircraft position errors in vertical e_z and horizontal e_{x-y} plane after transition into descent.

While the horizontal plane e_{x-y} theoretical position error of improved model (3) is negligible, the vertical plane e_z position error will be in the worst case almost equal to the reduced vertical separation minimum (RVSM) standard, in high–speed long–duration transition into descent, in tail–wind conditions (Fig. 4). However, as the vertical plane e_z position error is predictable and constant after transition into descent, the future aircraft descent trajectory is determinable and with corrections for the e_z, accurate as well.

3.2.2 Aircraft trajectory prediction errors

The descent trajectory prediction error of each model was determined by comparison of future trajectory predicted for the next 900 seconds (15 minutes) using each model (2) and (3) with the actual flight data recorded on commercial flight of Airbus A320 [10]. The ATC imposed break in actual flight which came after the TOD was used to foster trustworthiness of the methodology for the determination of the trajectory prediction error. The results of comparison are presented in Fig. 5, where point A indicates the TOD and B denotes the tropopause (ISA–1,24°C). At C the descent is interrupted, at D resumed, and at point E the descent speed regime has been altered from descent with the constant Mach number (0.78) to descent with the constant calibrated airspeed (280kt).

The trajectory prediction error of a primitive model of aircraft relative motion (2) clearly increases with the look-ahead time.The reason for such error is in the design of the primitive model of flight being ignorant to the variation of a true airspeed due to the static air temperature gradient in the troposphere. Within next 300 seconds (5 minutes) after the descent is resumed (at E), the trajectory prediction error of the primitive model exceeds 30% of the RVSM standard.

For the entire look-ahead time (15 minutes) of an aircraft future trajectory, predicted with the improved model of flight (3), its trajectory prediction error is stable in oscillations within ±16% of the RVSM standard. Being for the factor of at least 3 more accurate in trajectory prediction as the primitive model, the improved model promises greater reliability of conflict detection.

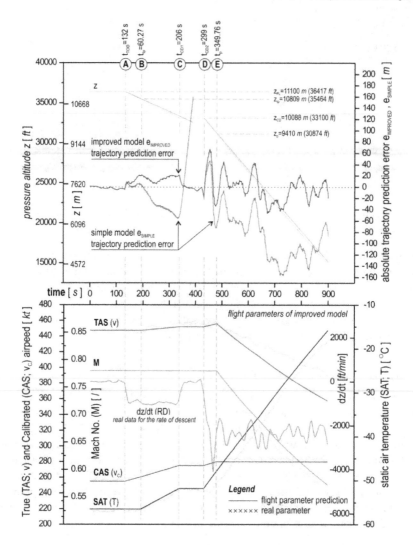

Figure 5. Trajectory prediction error of the primitive and improved model of flight compared to the flight data of real commercial flight.

3.2.3. Aircraft relative position error analysis

Initial separation $\mathbf{r}_{R.C}(\tau_{T/P}, \psi_R) = (x_{R.C}(\tau_{T/P}, \psi_R), y_{R.C}(\tau_{T/P}, \psi_R), z_R(\tau_{T/P}))$ between aircraft A2 and A1 crossing at relative heading ψ_R, is at the moment $\tau_{T/P}$ when A2 plans to initiate its descent at the TOD, critical (as shown in Fig. 6)if separation between them is lost after critical time tc while A2 descents:

$$x_{R.C}\left(\tau_{T/P},\psi_R\right)+\underbrace{\int_0^{t_C}F_x(t)dt}_{} \leq |r| + H\,\mathrm{tg}\,\theta_R(\sigma(h))$$
$$\underbrace{x_R\left(t_C,\psi_R\right)}$$

$$y_{R.C}\left(\tau_{T/P},\psi_R\right)+\underbrace{\int_0^{t_C}F_y(t)dt}_{} \leq |r| + H\,\mathrm{tg}\,\theta_R(\sigma(h)) \tag{22}$$
$$\underbrace{y_R\left(t_C,\psi_R\right)}$$

$$z_R\left(\tau_{T/P}\right)+\underbrace{\int_0^{t_C}F_z(t)dt}_{} \leq |H|$$
$$\underbrace{z_R\left(t_C\right)}$$

where r is separation minimum in a horizontal plane, H is separation minimum in a vertical direction while time functions $F_x(t)$, $F_y(t)$, and $F_z(t)$ are defined either by primitive (2; PRIM) or advanced model (3-19; ADV) of aircraft relative motion.

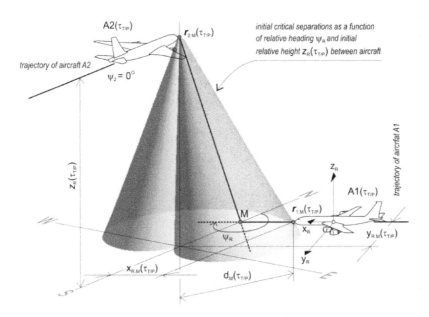

Figure 6. Critical initial separation between aircraft (collision in M is presented as special case of (22) where $r, H = 0$).

Figure 7 shows typical conflict detection error of the primitive model expressed with the error of relative distance between aircraft in close encounter situation in the horizontal plane ε_{dx} and ε_{dy} defined (according to (22) and Fig. 6) as:

$$\varepsilon_d\big|_{z_R(\tau_{T/P})} = \left[\underbrace{\left(x_{R.M}(\tau_{T/P}, \psi_R)\big|_{z_R(\tau_{T/P}):ADV} - x_{R.M}(\tau_{T/P}, \psi_R)\big|_{z_R(\tau_{T/P}):PRIM} \right)^2}_{\varepsilon_{dx}\big|_{z_R(\tau_{T/P})}} + \right.$$

$$\left. + \underbrace{\left(y_{R.M}(\tau_{T/P}, \psi_R)\big|_{z_R(\tau_{T/P}):ADV} - y_{R.M}(\tau_{T/P}, \psi_R)\big|_{z_R(\tau_{T/P}):PRIM} \right)^2}_{\varepsilon_{dy}\big|_{z_R(\tau_{T/P})}} \right]^{\frac{1}{2}} \qquad (23)$$

Imagine an aircraft A2 initiating the descent, then the vertical axis in Fig. 7 represent the relative height between A2 and intruder A1 below at the time of initiation of descent $z_R(\tau_{TOD})$. Note how fast the error with which the primitive model predicts future distance between aircraft in encounter incessantly increases after the descending aircraft starts to descent with the constant calibrated speed regime (after wasp-like contraction of a graph on Fig.7).

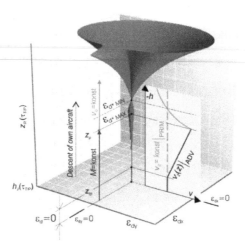

Figure 7. The relative distance between aircraft error of the primitive model of flight.

Investigation of the relative distance between aircraft error of the primitive model shows, as presented in Fig. 8, that the error ε_d will definitively (in any encounter situation) exceed r horizontal separation minimum (in Fig. 8 is represented by a cylinder). The peak values of relative distance between aircraft error ε_d are specific for head-on encounters ($140° < \psi_R < 220°$) and head winds relative to the descending aircraft A2 ($90° < \xi < 270°$; tail wind for the intruder A1). The error of relative distance between aircraft ε_d increases exponentially with the wind speed w (Fig. 8).Horizontal separation minimum r will be surpassed by relative horizontal distance error εd(23) sooner at smaller initial relative height between intruder A1 and descending aircraft A2 (10000ft@75kts of wind) in windy atmosphere and when the descending aircraft is faster than intruder.

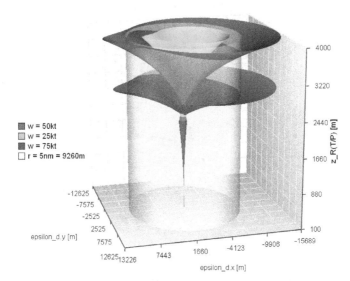

Figure 8. The relative distance between aircraft error of a primitive model exceeds the horizontal separation minimum.

Based on the critical initial separation between aircraft (22), the error of relative position between aircraft in the vertical direction εz depends upon the rate of descent and the speed regime change management of descending aircraft $dz(t)/dt\big|_2 = f(v_2(\sigma_2(z),\vartheta_S(z)),\psi_2,\theta_2(\sigma_2(z)),\vartheta_D)$ (4) and is defined as:

$$\varepsilon_z = \left(\frac{z_R(\tau_{T/P})}{F_z(t)}\bigg|_{ADV} - \frac{z_R(\tau_{T/P})}{F_z(t)}\bigg|_{PRIM} \right) \frac{dz(t)}{dt}\bigg|_2 \tag{24}$$

The analysis of the primitive model error of relative position between aircraft in vertical direction εz (24) is shown in Fig. 9. As long as the true airspeed increases (Fig. 6) in constant Mach number descend (6-12), the primitive model (2) is *slow* in defining the future vertical separation between descending aircraft A2 and intruder A1 below. Descending aircraft A2 will actually fly lower than predicted, and the actual vertical separation between aircraft will be smaller than predicted. After the speed regime change, the true airspeed will decrease (Fig. 6) in constant indicated airspeed descent (13-19), consequently the primitive model (2) is *to fast* in defining the future vertical separation between aircraft. Descending aircraft A2 will actually fly higher than predicted, and the actual vertical separation between aircraft will be greater than predicted. It is the constant indicated airspeed descent phase where the relative position between aircraft in vertical direction εz (24) of the primitive model increases exponentially and exceeds the vertical separation minimum (Fig. 9).

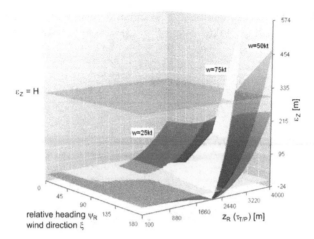

Figure 9. The primitive model error of relative position between aircraft in vertical direction (reduced vertical separation minimum is shown).

3.3. Reliability of conflict detection

The primitive model error of relative position between aircraft in vertical direction εz (24) will vary in a range of 10% of the RVSM standard for the conflict encounters up to 2000m (6500ft) below tropopause at moderate wind conditions (w = 26m/s (50kt)). However, after 5 minutes of descent and 3600m (12000ft) below tropopause, that is 70 km along the descent trajectory, the error of the primitive model will in vertical direction εz increase to 170% of the RVSM standard (Fig. 9).

The consequence is that, even before the relative horizontal distance error exceeds the horizontal separation minimum, the primitive model of relative motion becomes blind and unable to detect threatening conflict between aircraft. At the same time conflict alerts of the primitive model will be false resulting in the unnecessary conflict avoidance maneuvering which will actually be unsafe since it can lead into the undetectable conflict with yet another aircraft (domino effect).

The inability to detect loss of separation and erroneous conflict detection of the primitive model of aircraft relative flight (2) is addressed in Fig. 10. From (22) the minimal and maximal initial critical separations that will result in the loss of separation are determined and shown in Fig. 10. Clearly both, the area of undetected conflicts and the area of false conflict detection of the primitive model increase with increasing initial vertical separation between aircraft $z_R(\tau_{T/P})$ at the planned initiation of descent. Those areas increase exponentially for head-on encounters in windy conditions and if descending aircraft flies slower than the intruder. In case of moderate wind the sum of the area of false conflict detection and the area of undetected conflicts will increase to approximately 50% of the area of correct conflict detection in head-on encounter at initial vertical separation of 3000m (6500ft) if descending aircraft is 20% slower than intruder.

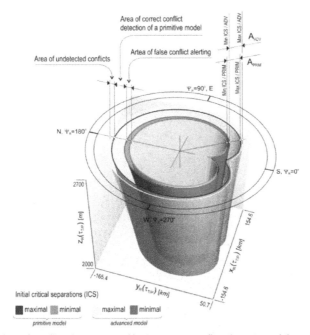

Figure 10. Inability of conflict detection and/or incorrect conflict detection of the primitive model in horizontal plane.

4. The infrastructure of autonomous flight

For the AASAS on-board aircraft to be based on the improved model of the aircraft relative motion (3) numerous information ofeach aircraft flight in the conflict detection zone has to be exchanged via Automatic Dependence Surveillance–Broadcast (ADS–B) as shown in Fig 11.

According to the improved model of the aircraft relative motion (3) those information (Fig. 11) are:

a. instantaneous flight parameters ζ_i: aircraft position, heading, speed regime and rate of climb or descent,

b. flight plan – i.e. future intentζ_P: set of each future navigational fix in place and time with corresponding flight parameter which will be altered at the fix, and

c. real time data on static ϑ_S (static air temperature at the atmospheric reference level) and dynamic ϑ_D state (wind speed and direction) of atmosphere (Fig. 12).

Since the trajectory prediction error pattern of the improved model of aircraft relative motion is non-increasing and stable with the absolute error less than the RVSM standard (Fig. 5), the AASAS on-board each aircraft will be, based on information exchanged, able to predict stable future four-dimensional traffic situation in the conflict detection zone around aircraft. In case of threatening conflict of separation loss the AASAS willaccurately define

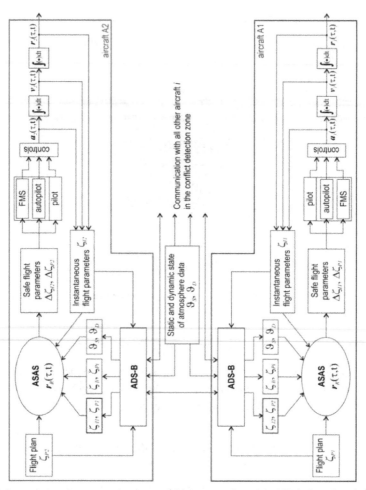

Figure 11. Information exchange requirements of the improved model of flight based Airborne Separation Assurance System.

the safe parameters $\Delta\varsigma_i$ and $\Delta\varsigma_P$ of conflict avoidance maneuveringfor the execution by the FMS/ autopilot or crew.

Investigation of the relative distance between aircraft error of the improved model as well as its trajectory prediction error revealed (Fig. 5), that the temperature at the atmospheric reference level is the improved model of aircraft relative motion accuracy most critical parameter. The necessary data on atmospheric conditions defined by (3) and their proposed format for up-link to the aircraft are presented in Fig. 12 and obtainable by the Mode-S transponder (for example).

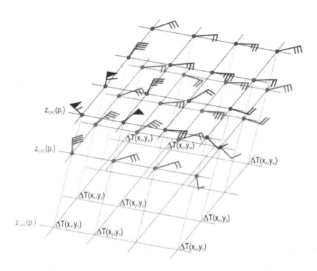

Figure 12. Proposed format of atmospheric conditions data available to each airborne aircraft.

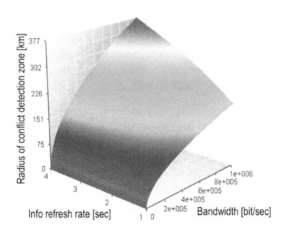

Figure 13. The feasible radius of conflict detection zone of the improved model of aircraft relative motion based ASAS.

The quantum of necessary information which has to be continuously and uninterruptedly exchanged between each aircraft aloft within radius of conflict detection zone of each other and with the ground systems providing them atmospheric conditions data impose concerns about ADS–B ability to exchange all those information. However, based on the assumption that the complete uncompressed data necessary comprise 1150 bits (comparison [6]) of exchanged massage among 30 aircraft per 100×100nautical miles (0.000875 aircraft per square km (Eurocontrol Performance Review Report 1999-2010)) on each of 28 flight levels from FL410 to FL150 (exaggerated aircraft density), and Universal Access Transceiver with the bandwidth of 1Mbit/s [6] is used, then the conflict detection zone with radius of up to 370 km (200 nautical miles) can be realized with the complete information refresh rate of 4 seconds. Relationships described and influence of information refresh rate and the data exchange bandwidth on the feasible radius of conflict detection zone are presented in Fig. 13.

5. Conclusion

Notwithstanding its many–sided complexity, the introduction of the AFA is inevitable, as ideas of unmanned cargo and passenger aircraft are emerging and the first UASs are already inexorably taken to the skies. The AFA technology development is applicable to the coming generation of aircraft and ATM systems with increasing automation anticipated to cope with increasing demand for airspace capacity.

Imminent increase in conflict encounter threatening aircraft transitioning to and from the AFA can be dispersed along the entire trajectory of aircraft with reduced severity of each remaining area of increase in conflicts with the introduction of descending or climbing transitions and AA reception/dispatch zone below the AFA where expected aircraft mix can be handled. The enabling technology is machine–based decision–making airborne AASAS and ground–based automation ASAS communicating by data–link. This AASAS should be capable of accurate conflict detection before descending aircraft exits the AFA or before ascending aircraft enters AFA. Otherwise the conflict encounters (loss of separation between aircraft) imminent in the AA Reception/Dispatch Zone of Fig. 1 are unavoidable.

The investigation of trajectory prediction error of a primitive model of aircraft relative motion clearly indicates the reason for the unstable prediction of future three-dimensional airborne traffic situation with existing (TCAS-like) technology and methodology. Furthermore, the conclusion can be drawn from the study that conflict detection and resolution is not safe and actually impossible for look-ahead time longer than 5 minutes in the airspace where aircraft are flying their trajectories in the vertical plane. Consequently, the primitive model of aircraft relative motion based airborne separation assurance cannot assure promptly and accurate conflict detection and therefore cannotfacilitate the AFA introduction.

The improved model of aircraft relative motion compared to its primitive pendant appears promising particularly because of the stability of its trajectory prediction error. This error might be less than described in the paper if the real reference atmospheric temperature is provided to the AASAS on-board each aircraft.

Plain proof is provided that the AFA and autonomous flight operations are feasible and basic–level AFA operational procedures are introduced. Crucial to the AFA introduction feasibility are technologies enabling: (a) sufficient bandwidth for reliable data–link communications, (b) capability to predict accurate and stable future 4D traffic situations with sufficient look–ahead time, (c) multi–factor analyses for real–time determination of safe transitioning trajectory including determination of the TOD, TC, and AA reception/dispatch zone collector airway selection, (d) adaptive airways structuring of AA, and (e) dynamic airspace sectorization.

Author details

Tone Magister and Franc Željko Županič
SLOVENIA CONTROL, Slovenian Air Navigation Services, Ltd, Ljubljana, Slovenia

Acknowledgement

The author is sincerely grateful to Cpt. Aleksander Sekirnik of Adria Airways for his unselfish support, advice and help.

6. References

[1] ACARE: "The Challenge of Security", in Strategic Research Agenda 1, Vol. 2, Advisory Council For Aeronautics Research in Europe, 2002.

[2] Beers, C.,Huisman, H.: "Transitions between free flight and managed airspace", 4th USA/Europe ATM R&D Seminar, Santa Fe, NM, USA, 2001.

[3] Bilimoria, K., Lee, H.: "Performance of air Traffic Conflicts for Free and Structured Routing", AIAA Guidance, Navigation, and Control Conference, Paper No. 2001- 4051, Montréal, Canada, 2001.

[4] Duong, V., et al, "Sector–Less Air Traffic Management," 5th USA/Europe ATM R&D Conf., Budapest, Hungary, 2003.

[5] Erzberger, H.: "The Automated Airspace Concept", 4th USA/Europe ATM R&D Seminar, Santa Fe, NM, USA, 2001.

[6] Hoekstra, J.M.: Designing for Safety – The Free Flight Air Traffic Management Concept, Ph.D. thesis, Technische Universiteit, Delft, The Netherlands, 2001.

[7] Kopardekar, P., Magyarits, S.: "Measurement and Prediction of Dynamic Density", 5th USA/Europe ATM R&D Seminar, Budapest, Hungary, 2003.

[8] Magister, T.: "The problem of mini-unmanned aerial vehicle non-segregated flight operations", *Traffic*, 2007, vol. 19, no. 6, pg. 381-386.

[9] Magister, T.: "Transition flight between the autonomous flight airspace and automated airspace", *Traffic*, 2008, vol. 20, no. 4, pg. 215-221.

[10] Magister, T.: "Long range aircraft trajectory prediction", *Traffic*, 2009, vol. 21, no. 5, pg. 311-318.

[11] Mondolini, S., Liang, D.: "Airspace Fractal Dimensions and Applications", 3th USA/Europe ATM R&D Seminar, Napoli, Italy, 2000.

Chapter 9

A Multi-Agent Approach for Designing Next Generation of Air Traffic Systems

José Miguel Canino, Juan Besada Portas, José Manuel Molina and Jesús García

Additional information is available at the end of the chapter

1. Introduction

Current implementation of new technologies for Communication, Navigation, Surveillance and Air Traffic Management (CNS/ATM systems) along with computational improvements on airborne and ground systems developed in the last two decades, point out the need for more strategic navigation and air-traffic control procedures based on four-dimensional (position plus time) trajectories. Moreover, the CNS/ATM infrastructure will help to achieve more shared real-time information among aircraft, airlines and air-traffic services providers (i.e. Air Traffic Control –ATC- providers, meteorological information providers and air space resources information providers). Then, general requirements for a next-generation of Air Traffic Management (ATM) system is that entities must share real-time information about aircraft state and intentions, air-space state and resources, meteorological conditions and forecast, etc. From this information coordinated and strategic actions should be carried out by them in order to achieve efficient and free of conflict 4D trajectories.

Several systems have been used on last years to help controllers and pilots to take decisions in a more strategic way. Some examples of these systems are OASIS [1], MAESTRO [2], COMPAS [3], CTAS, [4], etc. In addition, in the PHARE program [5], on-board trajectory predictions were used to inform the air traffic controllers about aircraft intentions. Results of the previous proposal showed potential advantages of a more strategic navigation and Air Traffic Control (ATC) based on 4D trajectories.

However new efforts are still necessary for achieving an ATM system characterized by a greater aircraft autonomy to select an optimal flight path and by a higher automation of air-ground coordination tasks. Thus, the Free Flight operational concept [6] suggests a future ATM where the aircraft are only restricted by global goals that must ensure safe and efficient air traffic flows. By other hand, the Trajectory Based Operations (TBO) [7] attempts to give more specific to the free flight proposal. In this case TBO concept proposes that coordination activities between ATM system elements are intended to achieve efficient and free of conflict 4D trajectories. In a hypothetical TBO scenario, aircraft will calculate their User Preference Trajectories (UPTs) taking into account air-space constraints. In addition, an ATC with the role of a central agent should drive air ground negotiation processes in order to provide free of conflict trajectories. Furthermore, under certain circumstances aircraft self-separation could be possible and air-air negotiation processes could be only supervised by the ATC. After a negotiation is performed the aircraft must fly the negotiated trajectories until a new set of trajectories were negotiated. Meanwhile, if emergence or contingence arises, a new negotiation process could be triggered.

In summary the new operational concepts propose:

- Four dimensional trajectory based operations defined by the aircraft position and time. These trajectories must suit the preferences of the aircraft while preserving the efficiency and safety of surrounding air traffic flow.
- Accessibility and distribution of updated data among all entities involved in flight operations.
- A more distributed reallocation of the roles of aircraft and services of air traffic control to achieve their respective goals, in contrast with the current scheme of responsibilities characterized by a ground-centralized monitoring and air traffic separation activities.

Application of these operational concepts by 2020+ is the key target of current research initiatives such as SESAR (Single European Sky ATM Research) and Next-Gen (Next Generation Air Transportation System) [8-9]. These research programs show that the set of activities aimed to validate procedures and technologies in order to implement the cited paradigm is diverse and extensive. To identify the interdependence of such activities, we propose a framework that classifies them according to a sequential time process (see Fig. 1). The first two groups of activities are referred to the analysis of the potential of CNS/ATM systems and as well as feasible operational concepts: i.e. Free Flight, TBO, etc. Proposals of operational concepts consist of generic specifications that requires of concrete procedures for conducting operations navigation and air-traffic control. Parallel to procedures design, on-board and ground systems and underlying mathematical models have to be also developed. Then, initial design of procedures and their associated systems can be considered as an iterative process that needs to be validated by means of analytical simulation. This process is a key issue as a previous stage to the Human-In-The-Loop (HITL) simulations and flight tests. HITL simulations and flight trials allows defining specific standards (e.g Procedures for Air Navigation Services –PANS- or for Required Performance Navigation –RNP-) for the actual implementation of the operational concept.

Several attempts for developing simulation and design analysis tools have been proposed [10-16]. Results of these studies show that it will be necessary more detailed and structured conceptual models to give support to analytical simulation tools to address the paradigm shift in the ATM procedures. Moreover, the close interdependence between coordination procedures, systems to execute them and their underlying mathematical models must be clearly set out in the conceptual model.

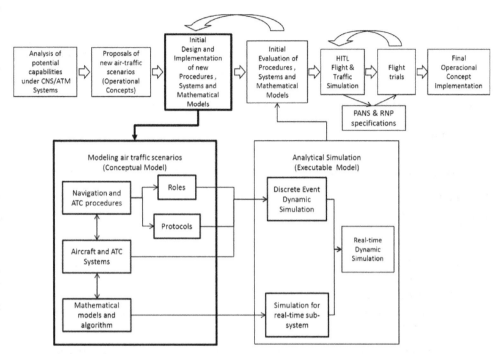

Figure 1. Research and development activities for new operational concepts

Fortunately, current state of the art of Multi-agent technologies provides methodological approaches and tools to develop such models. Then, the main goal of this chapter consists in illustrating how recent proposals in agent-oriented methodologies are a suitable approach for analysing and modelling new air traffic scenarios in order to obtain such conceptual models.

The chapter begins summarizing main contributions on modelling and simulations of future ATM. This review shows that new simulation platforms should be based on more detailed and structured conceptual models. In additions, it suggests benefices of multi-agent methodologies for approaching this problem. In section 3, multi-agent approaches for modelling several aspects of cited ATM are analysed. After a brief description of particularities of the multi-agent theory, a review of its most recent applications within the air traffic scope is presented. This exploration highlights the need of taking advantage of modern multi-agent technologies for developing robust and modular conceptual models of the new ATM. Section 4 presents the multi-agent methodologies as useful tools for analysing and modelling this ATM in order to facilitate the design and take key decisions for the implementation of new procedures. From them, one of the most recent agent methodologies, named Prometheus, has been selected to integrate the system specifications, inter-agent coordination protocols and agent inner processes as components of an ATM conceptual model, described in section 5. In order to simplify the applicability of the mentioned methodological approach, it has been applied for analysing and modelling a particular air traffic scenario of arrival air traffic operations. In this case we will consider this particular ATM model as an Air Traffic System (ATS). Moreover the model is focused on the inner architecture of an ATC System. Although these simplifications, the obtained model under this methodological approach can be extended for adding new gate-to-gate air traffic scenarios, coordination procedures as well as improved versions of the technical support to execute mentioned procedures. Section 6 resents guidelines for a software implementation and results of an illustrative example of current implementation state. Finally, conclusions are presented in section 7.

2. Simulation and design analysis tools for new air-traffic concepts

The simulation of air traffic scenarios with different levels of fidelity is the mainstay of the methodology used widely by the scientific community to develop new operational concepts [8-9]. Real-time simulations are, in general, HITL simulations intended for human factors evaluation while navigation and/or traffic control procedures are performed. Fast-time simulations are centred on the analysis of several issues no specifically focused on human factors: mathematical models, algorithms, negotiation and/or decision-making processes, quality of service measures, etc. Obviously real-time simulation platforms are able to carry out fast-time simulations by including computer models for performing tasks assigned to the human element.

Some of the multi-aircraft simulators currently used for air traffic research purposes are: NLR's Air Traffic Control Research Simulator (NARSIM)) [10], Pseudo Aircraft System (PAS) [11], Target Generation Facility (TGF [12], ATC Interactive for the future of air traffic control[13] and Multi-aircraft Control System (MACS) [14]. Three main consequences can be derived from the analysis of previous works. First, more detailed proposals are necessary for supporting automated air-ground coordination processes. Second, modelling such automatic coordination procedures system requires a parallel specification of their associated systems (i.e. user interfaces, mathematical models and algorithm for making decisions, etc). Third, new conceptual models and simulation frameworks require a high modularity and scalability for making possible the progressive incorporation of new or modified procedures and their associated systems as they are designed or evaluated.

In the context of modelling such complex distributed systems, Multi-Agent Systems (MAS) theory gives natural solutions [17-18]. This theory provides a suitable framework to analyse and model the organization of a set of autonomous ATM entities that coordinate and negotiate their actions in order to achieve their respective goals.

3. Multiagent approaches for future air traffic scenarios

The dynamic nature of air traffic and its geographical and functional distribution have attracted the attention of agent researchers since the last decade. For example, Optimal Aircraft Sequencing using Intelligent Scheduling (OASIS) [1] is a system used at the airport of Sydney to help air traffic controllers on arrival and approach air traffic sequencing operations. However, most of agent-based contributions are oriented to developing very basic aspects of the ATS system. The proposals can be classified into three categories:

i.   Analysis of negotiation patterns between agents in free flight air traffic scenarios. Within this category, Wangerman and Stengel analyse the dynamic behaviour of aircraft in the airspace as an intelligent system (Intelligent Aircraft/ Airspace System or IAAS) [19-21]. This system consists of three types of agents, airlines, aircraft and traffic managers, and the dynamic behaviour of the agents of IAAS is analysed from the perspective of a distributed approach. In this context, the principles of negotiations are a way to implement distributed iterative optimization of IAAS operations. In [22] Harpert et al. propose an agent-oriented model of an ATM in a free flight and a model of distributed decision making for the resolution of conflicts in the ATM. This proposal includes a distributed optimization scheme in which the agents generate and evaluate proposals options that best suit their preferences on a utility function and a multi-attribute decision tree scheme. Besides, the declarative capabilities of an intelligent agent can be modelled by expert systems [23-25]. To set the ground rules of the expert system, tasks were classified in [21] into four groups: emergency tasks, tasks of a specific mode of operation, negotiation tasks and routine tasks.

ii. Avionics systems for autonomous operations in distributed air traffic management scenarios. Works in this category propose designing avionics systems based on multi-agent systems. The proposed developments basically consist of systems for automatic conflict resolution, automatic warnings and recommendations to the crew [26-28]. In [26] authors propose a design of intelligent traffic agent developed to detect and solve conflicts on board in a free flight environment. An extension of previous work presented in [27] proposes a design of an executive agent that resolves conflicts for both the traffic and bad weather areas. In this case an agent's hierarchical architecture is suggested for making decisions based on the information produced by a traffic agent proposed by [26] and a weather agent. Finally, in [29] capabilities required of future flight management systems (FMS) in the cabin were characterized by means of agent-oriented analysis of the air navigation and arrival operations into a distributed environment. This analysis was later extended to define capabilities required to carry out automated arrivals and departures at uncontrolled airports [30].

iii. Simulation systems for the analysis of advanced air traffic concepts. This group includes several simulation platforms used for the design and validation of procedures and systems proposed for next generation of air traffic scenarios. In [15, 17] a multi-aircraft control platform is proposed to increase the realism and flexibility of HITL simulations. Functional descriptions of pilot and ATC perspectives within this platform are presented in [15]. In [17] the ATC agent model is analysed in order to identify its roles and responsibilities in future ATM systems. Another development of functional architecture airspace for an Airspace Concept Evaluation System (ACES) is presented in [16]. In a later work an agent-oriented model of the CNS/ATM infrastructure of ACES was proposed in [31]. Finally, in [32] a design of an experimental air traffic simulator implemented as a Java environment SMA is presented.

The proposals based on multi-agent systems presented above cover a wide spectrum of issues related to air traffic operations in a free flight environment. However it is clear that effective implementation of the new operational concepts involved in the future scenario still requires more structured models that take into account the tight relationship between procedures and systems to support them. In addition, as it was explained before, the model should be scalable enough to allow a progressive integration of the following elements into the model:

- Operating procedures for ATC and aircraft, specifying: (i) roles and tasks assigned to ATC and flight crew and (ii) coordination rules and negotiation protocols among the involved agents
- ATC and on-board functionalities to help to execute such procedures at several automation levels.
- Underlying mathematical models and algorithms to give support to previous functionalities.
- High level languages that allow for accurate intercommunication between aircraft and ground systems.

Fortunately, methodologies for developing multi-agent systems have reached a noticeable degree of maturity in recent years, becoming an invaluable tool to achieve a comprehensive analysis and modelling of complex scenarios. Thus, the analysis and modeling of interactions in terms of coordination and negotiation strategies between agents provides useful guideline for developing new schemes for developing automated ATC and navigation procedures. In turn, the study of the agents' behaviour and their internal architecture for the mentioned coordination processes provides a more precise identification of the functionalities required by on-board and ground systems to execute mentioned these procedures.

4. Current methodologies and tools for analysing and designing MASs

Agent methodologies provide with a set of guidelines to facilitate the development of multi-agent systems over several stages since the initial draft of idea until the final detailed design. In this way, current multi-agent technology provides practical and formal methodologies to analyse and design, in a structured and consistent manner, the following issues: (i) roles and functionalities of autonomous entities (agents) that take part in an operational scenario, (ii) interactions between agents (or agent protocols) and (iii) inner architecture and dynamic behaviour (processes) of agents. Besides several agent platforms have been proposed as middleware tools for translation the conceptual model into an executable model.

The design of MAS requires not only new models but also the identification of the software abstractions, since this paradigm introduces a higher abstraction level when compared to traditional approaches. They may be used by software developers to more naturally understand, model and develop an important class of complex distributed systems. The key aspects of problems being modelled under a MAS methodology are: establishing a set of coarse-grained computational systems (agents) and interaction mechanisms for a goal to obtain in the system that maximizes some global quality measure, assuming a certain organizational structure which can be assumed to keep fixed (agents have certain roles and abilities that do not change in time).

Some of the main agent-oriented methodologies are MASCommonKADS [33], Tropos [34], Zeus [35], MaSE [36], GAIA [37], INGENIAS [38]. In [30-41] a comparative analysis of the main methodologies is presented. MAS-CommonKADS is a methodology for knowledge-based system that defines different models (agent model, task model, organizational model etc.) in the system life-cycle using oriented-object techniques and protocol engineering techniques. Tropos is a requirement-based methodology; Zeus provides an agent platform which facilitates the rapid development of collaborative agent applications; MaSE is an object-oriented methodology that considers agents as objects with capabilities of coordination and conversation with automatic generation of code and Unified Modelling Language (UML) notation; Gaia is intended to go systematically from a statement of requirements to a design sufficiently detailed for implementation; and INGENIAS proposes a language for multi-agent system specification and its integration in the lifecycle, as well as it provides a collection of tools for modelling, verifying and generate agents' code.

A descriptive analysis these methodologies are beyond of these chapter goals. However after a previous study we have selected Prometheus methodology as the most suitable one. Prometheus agent-oriented well-established methodology has been selected to provide guidelines to develop the mentioned multi-agent system [42]. We argue that Prometheus suits well for solving our problem due to: (i) the highly detailed guidelines for defining the initial system specification, (ii) the modularity of the agent's internal architecture around the concept of capability (providing a direct correspondence between capabilities and functionalities of airborne and ground systems), (iii) the easy translation from the conceptual model into an executable model by means of current agent platforms that provides software infrastructure as it will be explained later on.

Prometheus methodology covers the entire process of design and implementation of intelligent agents. It includes three phases (see Figure 2): system specification, architecture design and detailed design [42].

System Specification stage defines the objectives or goals of the system. Goals help to identify functionality required to achieve them, as well as a description of the interface between the system and its environment in terms of inputs (*Perceptions*) and outputs (*actions*) of the system. The identification and refinement of the objectives are carried out, in an iterative manner, from the definition of different use case scenarios. Scenarios illustrate the operation mode of said system. The concept of scenario (or use case scenario) comes from the object-oriented software methodologies, but with a slightly adapted structure that provides a more integrated than the mere analysis of the isolated system.

Later on, in the Architecture Design phase, the description of the system structure is represented by means of diagrams that identify the agents of the systems and their interactions in terms of communication messages and protocols. Protocols represent specific communication schemes. They can be modelled using Agent Unified Modelling Language (AUML) that describes the interactions of agents in different scenarios of use cases.

Finally, the Detailed Design phase consists of designing internal processes carried out by each agent and an inner architecture that describes how these processes are organized and implemented. Prometheus proposes implementing agent tasks by means a set of different plans that are triggered when determinate events occurs. Plans are grouped into several groups associated to the execution of specific tasks. A group of plans as well data used or produced by them constitutes an agent capability. Then focus of this stage is to define capabilities, internal events, plans and detailed data structures. In this way capabilities are modules within the agent that use or provide related data types. Capabilities can be nested within other skills so that in the detailed design, the agent will have an arbitrary number of layers in an understandable complexity at each level. Thus, capabilities can be constituted by other sub-capacities or, at lowest level, by plans, events and data. The plans set out the set of tasks performed to achieve a particular purpose. They are also triggered by certain events (internal or external messages, perceptions, etc.). As a result of this stage are general diagrams of each of the agents (which show higher level capabilities of the agent), charts of capabilities, descriptors detailed plans and data descriptors.

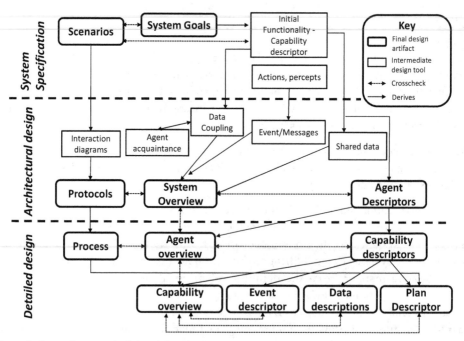

Figure 2. Prometheus methodology phases [42]

Moreover, the tool Prometheus Design Tool (PDT) facilitates the tasks of the developer over the previous stages to provide information about possible inconsistencies in the design [43]. In addition, several software tools have been developed in recent years for software implementation of system multi-agents. One of the most extended is the Java *Agent DEvelopment Framework (JADE)* Platform [44]. JADE simplifies the implementation of multi-agent systems through a middleware that provides several resources through a set of library classes aimed to:

a. Implement agent's tasks into a several JAVA classes named behaviours.
b. Provide agent intercommunication that complies with the *FIPA (Foundation for Intelligent physical Agents)* specifications [45].
c. Supply services for create and manage cycle of live of agents as well are their services into the multi-agent system.

JADE behaviours can be classified onto simple behaviours and composite behaviours. In turn, simple behaviours can be classified as:

a. One-shot behaviour, an atomic task to be carried out once, used here for initialization tasks;
b. Cyclic behaviour, which is iterated while exists, such as messages listening and processing;
c. Waker behaviour, or a one-shot behaviour invoked after a certain time; and
d. Ticker behaviour or a cyclic behaviour which performs a series of instructions executed keeping a certain fixed time, used in the platform for simulation numeric computation and graphical output.

Composite behaviours are three:

a. FSMBehaviour that consists of a class that allow defining a Finite State Machine by means sub-behaviours, where each of them represents an machine state
b. SequentialBehaviour that executes its sub-behaviours in a sequential way, and
c. ParalellBehaviour that executes their sub-behaviours concurrently and ends when a certain condition is satisfied (for one, several or all of them). In this way, agents are able to concurrently to carry out different tasks and to keep simultaneous conversations.

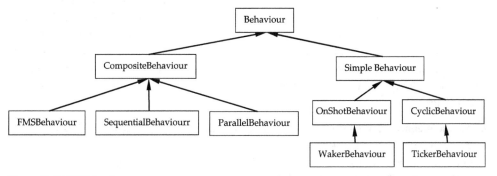

Figure 3. JADE Behaviours

5. Appling Prometheus methodology for designing an arrival TBO conceptual model

As explained in the previous section, Prometheus methodology carries out an iterative process on three phases: specification system, architecture design and detailed design. Each of these phases provides guidelines for designing several model components. These components produce a hierarchical structuring mechanism which makes possible a model description at multiple levels of abstraction [19]. In addition, the structured nature of design components facilitates crosschecking for completeness and consistency of the model in each design phase.

5.1. System specification

In this phase, goals of our ATS model are identified. In turns goals are captured from a set of *scenarios* that illustrate essential aspects of the system operation. Scenarios and goals help to recognize initial system functionalities and to examine the system-environment interface in terms of inputs *(Perceptions)* and outputs *(actions)* [45]. Thus, scenarios are use cases that contain a sequence of steps each of them relating to a goal, an action, a perception or another scenario.

For outlining the mentioned scenarios, a generic automated air traffic scenario was considered as a distributed process where several autonomous and proactive entities (agents) plan and execute a set of coordinated tasks to provide arrival and approach free of conflict 4D trajectory. This operational scenario is particularly critical in arrival air traffic operations due to the high variability of the speed, heading and altitudes that could affect to the degree of predictability of several converging trajectories. Moreover, guidelines from scenario proposed in DAG-TM (CE-11) project [46] have been taken into account. According to referred guidelines the flight crew: *(i)* could negotiate arrival preferred trajectories with the ATC; *(ii)* is responsible for maintaining longitudinal spacing between consecutive aircraft once a trajectory (o constraints) has been assigned.

In this operational scenario, the following agents have been identified: Aircraft, Air Traffic Control (ATC), Meteorological Service Provider (MPS), Airspace Resources Provider (ASP) and Airline Operational Control (AOC). In addition several ATC agents could be defined in order to coordinate arrival ATC tasks with the ATC en-route or departure ones. MSP, ASP and AOC agents' functionalities have been used to define the information required by the ATC and aircraft agents as well as essential protocols to accomplish this information.

Use case scenarios have been selected and organized taking into account perspective that each agent have about the generic air traffic scenario. Then, five root scenarios have been defined: *(i)* Manage Aircraft, *(ii)* Manage ATC, *(iii)* Manage Airline Operational Control, *(iv)* Provide Airspace Resources *(v)* Provide Weather Information.

Tasks of each one of above scenarios have been grouped into new sub-scenarios and so on. Figure 4 depicts a list of the most significant sub-scenarios deployed from the previous one. In particular *Manage navigation procedure* scenario and *Manage ATC* scenario are developed until the lowest level. In addition, this scenario architecture shows that air-ground negotiation processes are contained into specific air-ground negotiation scenarios which are shared by *Manage Aircraft* and *Manage ATC* scenarios.

To illustrate how scenarios can be deployed we will focus on the *Manage ATC scenario*. This scenario contains the following four scenarios:

- *Update ATC environmental information scenario*, which covers associated processes to collect information about status and intentions of aircraft, airspace resources (including restricted flight areas), weather conditions, etc.
- *Manage ATC procedures scenario*. It includes the processes related to maintaining the separation of aircraft to achieve an efficient traffic flow.
- *Manage on board surveillance scenario*. This scenario contains tasks for monitoring air traffic. These tasks are aimed at identifying aircraft trajectory deviations and potential conflicts with other aircraft or obstacles. It also provides viable solutions for correcting these anomalies and events for triggering specific processes in order to implement the solutions mentioned above.

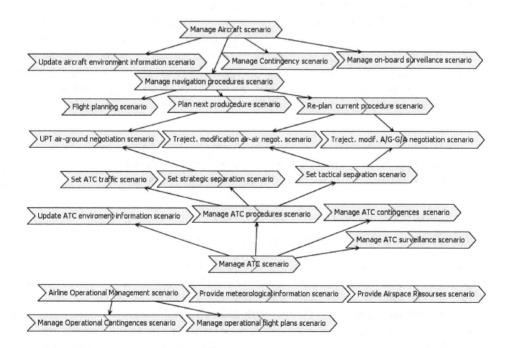

Figure 4. Architecture of the main scenarios for Trajectory Based Operations

- *Manage contingences scenario* that includes tasks for analysing air traffic contingencies and circumstances in which they occur (e.g. aircraft malfunctions, on-board contingences, etc.). It also includes decision-making processes to determine actions to be carried out regarding the management of traffic control procedures.

Focusing on the *Manage ATC procedures scenario*, the next three scenarios are deployed:

- *Set ATC traffic* involves actions for *receiving* or transferring air traffic from or to other adjacent ATC agents.
- *Set strategic separation*. This scenario contains tasks for planning aircraft trajectories and assigning them by means a negotiation process. *Therefore* it is a key scenario for modelling automated procedures for TBOs and it should contain several negotiations sub-scenarios.
- *Set tactical separation*. This scenario contains tasks for modifying current flight trajectories when unforeseen contingencies arise. The tasks performed in this scenario are twofold: (i) to provide specific instructions for activating protocols aimed at aircraft separation in extreme situations of short-range conflicts and (ii) to authorize and supervise air-air negotiation for self-separation when separation responsibility has been delegated on the aircraft.

From the above scenario architecture a goals tree can be obtained. Lowest level goals help to identify functionalities and processes that the agent has to carry out to achieve them. For example, Figure 5 shows a particular set or goals that results from the *ATC manage* scenario. On it, the goal *UPT air-ground negotiation* consists of several sub-goals such as generate proposals for aircraft, evaluate proposals from aircraft or establish and an air-ground communication protocol for negotiate mentioned proposals and counterproposals. In addition, functionalities help to identify actions, perceps and data used or generated by the agents. Then, for the ATC agent the following perceptions can be identified:

- Perceptions form external sensors: data from radar systems, WAN receivers, etc.
- Perceptions from human-machine interface: menus and inputs options.
- Actions: display traffic data and ATC procedure state data on screams.

5.2. Architecture design and negotiation protocols

In the architecture design phase the following aspects of the overall system are defined:

- The system *overview diagram*. This diagram represents the static structure of the system, tying agents and main data used by them as well as their perceptions and actions. Furthermore communication interactions between agents are considered.
- The set of *interaction protocols* that capture timing of communication of related messages between agents. These protocols are derived from the scenarios defined in the specification phase protocols and, therefore, they describe the system dynamic behaviour. Then, they have been depicted using an AUML notation [47] .

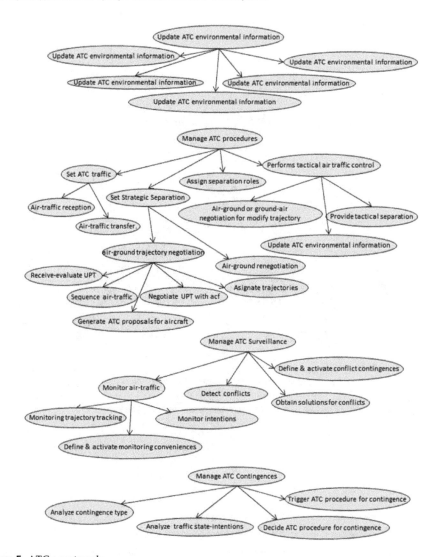

Figure 5. ATC agent goals

Figure 6 shows a simplified representation of the system overview diagram. The simplification consists on representing only the main actions and perceptions of the ATC and aircraft agents as well as the main communication protocols. In a more complex system overview diagram, all these elements should be signed for all the agents.

After identifying the interaction protocols in the system overview diagram, protocols are designed. For the ATC and aircraft agents, protocols are aimed to: (i) improve agents' knowledge base about the environment and/or the other agent's intentions, (i) negotiate trajectories that could be in conflicts.

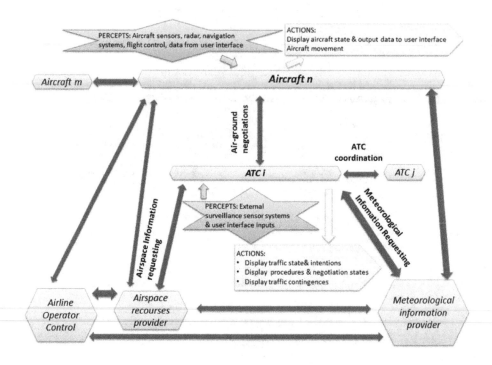

Figure 6. Simplified architecture overview

For a better understanding of automated air-ground coordination aspects, we will focus on describing a proposed air-ground negotiation protocol (see Figure 7). This protocol represents the core of the both the ATC strategic planning tasks and the aircraft navigation planning tasks. Although new scheme of negotiation can be defined from this design, all of them will use similar functionalities to evaluate proposals and generate counter-proposals. Therefore, this protocol and its associated functionalities provide guidelines and specification enough for developing new aircraft and ATC coordination procedures.

In Figure 7, the on board computation processes are represented on the left side of the aircraft agent lifeline. On the right side of the ATC agent lifeline we can observe computation performed by ground systems. Moreover, on the right side of this ground computation system, a new lifeline for other aircraft agents is showed.

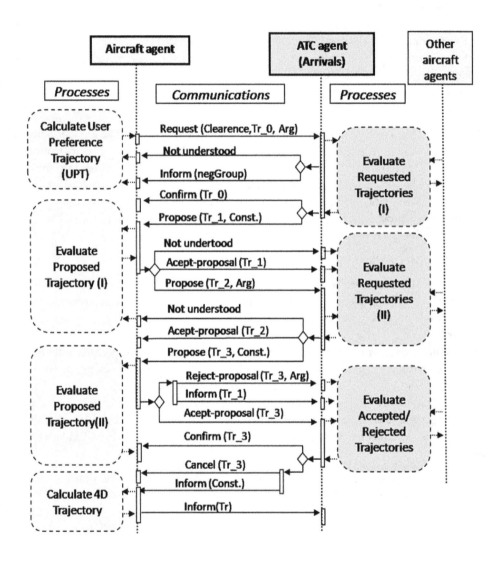

Figure 7. Air-ground Negotiation Protocol

The proposed arrival negotiation protocol can be divided into two phases that are described next:

a.   Phase 1: Before reaching Time Limit for Requesting Trajectory (TLRT)

In this phase aircraft calculates their preferred 4D Trajectory (Traj_0). To perform this computation, each aircraft agent uses the available information about meteorological conditions and arrival routes. This information is obtained by means of a previous communication procedure (no represented in Fig. 7) with the Meteorological Information Provider agent and the Air-space Resource Provider agent. Once the 4D trajectory is calculated, the aircraft requires clearance to the ATC to execute it. In this case, the TLRT represents a deadline time for requesting mentioned clearance.

The ATC agent receives *Request messages* from different aircraft that are periodically processed in-batch. After receiving these messages, the ATC evaluate if requested trajectories are free of conflicts. As a result, the ATC could confirm the trajectory initially preferred by the aircraft or it could propose new constrains for a new one *(Traj_1)*. If the aircraft is agreed with previous information, it sends a corresponding message and the communication process finalizes. But if the aircraft wish to flight an alternative trajectory (i.e. a faster one), the negotiation protocols continues in a second phase.

b.   Phase 2: Call For Proposal

In this phase, the aircraft makes a second counter-proposal in order to improve the previous ATC proposal, arguing reasons for it (for example certain operational contingences). These kinds of proposals *(Traj_2)* will be evaluated by the ATC. Those one that can be feasible will be accepted. In other case, the ATC will perform a new proposal *(Traj_3)* that the aircraft in turn can refuse or accept. When an aircraft refuses mentioned ATC proposal, it will have to select one that satisfy previous ATC constrains. But if the aircraft accepts cited ATC proposal, it will have to wait an ATC confirmation message before implement such proposal. This confirmation is necessary due to the ATC has to analyze air traffic state after receiving several aircraft messages accepting or refusing *Traj_3* proposals. Then the protocol ends with aircraft messages informing about details of the last cleared and accepted trajectory.

Finally note that the software implementation of messages used in this protocol can be performed by a normalized FIPA support [45].

5.3. Detailed design

Finally, in the detailed design phase, the dynamic behaviour and the internal agent architecture are projected. The dynamic behaviour is described by a set of processes that agents carry out when they interact or make decisions. The internal agent architecture is represent by means an agent overview diagram that shows how these processes are organized.

Processes are represented by a flow diagram that links protocol messages with internal functionalities that evaluate and generate proposals. Notation used for depicting processes is showed in Figure 8. This notation is slightly different to the UML notation so that, instead of focusing on the activities it is focused on communications. Besides, we extend notation proposed by Prometheus methodology in order to include information about the different states of the air-ground negotiation protocol. These negotiating states are intended to enable automated negotiation processes whose evolution can be understood in supervisory tasks of pilots and controllers. Figure 9 shows the process carried out by the ATC while the air-ground negotiation protocol, previously presented, is in progress.

Agent processes like the described above can be implemented by means of plans. Then plans contain a set of instructions in order to: (i) carry out computations, (ii) take decisions (iii) generate or receive messages and new events. Moreover plans are to be triggered by specific events such as arrival messages or events generated by other plans.

The agent diagram overview consists of an agent architecture representation that indicates how all these plans are organized. Therefore, it shows interaction between plans, shared data and events. In addition Prometheus methodology proposes to organize plans that share similar functionalities and data into capabilities. Figure 10 represents Prometheus notation for representing elements of the agent overview diagram. Then, Figure 11 represents the ATC agent architecture diagram overview. On it, main capabilities of the ATC agent are depicted together with data used or produced, agent inner events and communication messages.

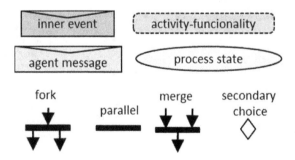

Figure 8. Notation used wihin agent processes

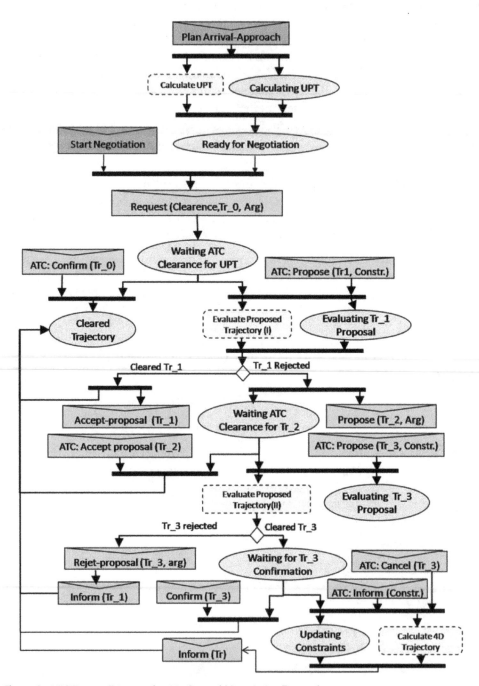

Figure 9. ATC Process Diagram for Air-Ground Negotiation Protocols

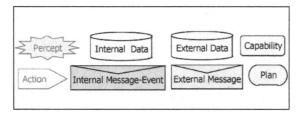

Figure 10. Prometheus notatino used in agent and capability overview diagrams

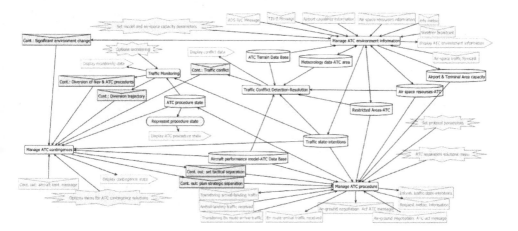

Figure 11. ATC agent architecture

Taking into account scenarios, functionalities, processes, events and data established in previous phases, plans has been grouped into the following capabilities:

- *Manage ATC Environment Information:* This capability is associated with the goal of maintaining an updated on-board environmental knowledge. Plans for this capability capture information about weather forecast, restricted areas, air space recourses (e.g. available arrival routes and gateways), air space contingency events concerning to significant environmental changes and aircraft contingence events.

- *Traffic Monitoring:* This capability checks the state and intentions of the aircraft according to air-ground agreements previously negotiated. In case of divergences, an air-traffic contingence event is generated informing about it.

- *Traffic Conflict Detection-Resolution:* As its name suggests, it is responsible for detecting conflicts with other aircraft or obstacles (terrain, adverse weather areas, etc.). It also provides a set of ranked proposals for conflict resolutions. Furthermore, proposals are negotiated and/or implemented by means of other capabilities. To achieve above goals, plans of this capability are grouped into two sub-capabilities: *(i) Conflict Detection Capability and (ii) Initial Conflict Solution Capability.*

Conflict Detection Capability contains plans to implement algorithms for conflict detection. Therefore, it can be constituted by several plans each of them contains a specific model to detect short, medium and long term conflicts. Plan inputs are data about predicted trajectory, restricted areas, surrounding traffic state and intentions. Plans of this capability are triggered by events generated by plans of other capabilities that perform surveillance tasks as well as testing tasks within the trajectory planning processes. Conflict data calculated by previous plans are used by a specific plan to obtain a detailed conflict description and to generate conflict events.

Initial Conflict Solution Capability uses several inner plans to supply solutions according conflict data input. Results of these plans are used by other ones that generate conflict contingence events. Then events also contain associated information about feasible conflict solutions.

- *Manage ATC Contingency.* This capability deal with deciding which kind of ATC procedural tasks are to be carrying out according to the information contained on received contingency events. To make decisions, plans of this capability take also into account current states and intentions of aircraft traffic. Information about the procedural tasks that have to be performed are included into contingency output events that will trigger specific plans to execute such tasks. These plans are grouped into the ATC Procedures Management capability that is described next.
- *Manage ATC Procedures.* Plans of this capability carry out strategic and tactical actions aimed to maintain aircraft separation. These plans are grouped into the next for sub-capabilities:
  - *Implementation of ATC procedures.* This capability has a first plan that take into account air traffic conditions to generate events that trigger plans for: (i) ATC coordination in order to receipt o transfer air traffic, (ii) planning and assigning trajectories, (iii) establishing point for initiate air-ground negotiations and (iv) assuming or delegating aircraft separation responsibilities.
  - *Strategic Separation.* This capability is modelled through two basic plans. One of them drives processes of trajectories negotiation. The second one manages other supplemental plans that implement re-negotiation processes of trajectories previously assigned to a group of aircraft and pending of execution when a contingency arise.
  - *Tactical Separation.* Plans of this capability manage processes triggered by contingencies that require this type of action (e.g. separation loss contingency when air traffic flows converge). Obviously, in the context of TBO, tactical actions should be reduced to: (i) delegate or regain the separation control role depending as a function of the air traffic state and other contingences and (ii) enable separation control protocols in extreme short-range conflicts.
  - *ATC Coordination.* This capability includes plans to coordinate for air traffic transferring between adjacent ATC units. Events that trigger these plans come from the sub-capacity ATC procedure execution. The detailed design of this capability includes a specific plan to implement ATC coordination protocols with other adjacent units.

6. Implementation

Descriptors and diagrams of the components in the described conceptual model contain all the necessary information to carry out implementation. However, not all the components obtained in the three phases of the methodology have to be implemented. The executable model consists of the entities that have been developed in the detailed design phase (i.e. agents, capabilities, plans, data, events and messages).

As it was explained in section 4, Prometheus methodology provides a full life-cycle support tool (PDT tool) to develop multi-agent systems. Current version of PDT provides support for: *(i)* designing most the design artefacts within the Prometheus methodology, *(ii)* cross-checking for consistency and completeness for the conceptual model, *(iii)* automatic generation of skeleton code in JACK agent-oriented programming language [48].

The conceptual model detailed above is currently at implementation phase. Although facilities of automatic code generation of PDT, we have opted for using the JADE Platform [39], cited in section 4, due to: *(i)* it is one of most extended multi-agent platforms and, *(ii)* it provides the FIPA standards [49] infrastructure for inter-agent communications and for managing software agents distributed across multiples hosts. As it was explained, architecture of JADE agent is built upon the behaviour concept rather than a plans-based architecture. Then agent plans can be generally implemented into JADE behaviours in a quite straightforward way.

On the other hand, continuous simulation requires, in nature, implementing the aircraft dynamic over a continuous-time model. It is essential to carry out real-time and human-in-the-loop simulations in order to analyse in detail and validate the design accordingly to the expected global behaviour. Also it is suitable for fast analytical simulations intended for preliminary designing and evaluation of cockpit systems and underlying mathematical models and algorithms (e.g. for 4D-trajectory guidance, conflicts detection and resolution, etc.). However, while the detailed models are not available, the proposed conceptual model enables discrete event simulation. In this case, events can be generated by random functions implemented within capabilities plans representing underlying models as *black boxes*. In this way, for an initial implementation phase, random functions to generate events are implemented into agent plans. In a later phase, the executable model can be refined when functions are replaced by specific underlying models as they are developed.

6.1. ATC model implementation under JADE platform

Figure 12 illustrates an adaptation of the ATC agent capacities-based architecture to a behaviours-based one using JADE behaviours described in section 4. Each one of the agent capabilities has been defined as behaviours that run in a parallel way. Previous behaviours could be progressively broken down into new behaviours, so that, at lower-level, behaviours correspond to plans of the conceptual model.

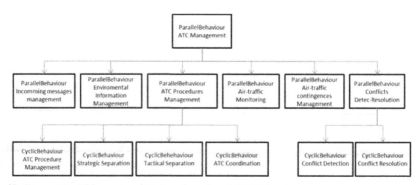

Figure 12. Figure 12. ATC agent architecture based on JADE Behaviours

6.2. Dynamic prototype example: An experimental air traffic simulator

As example of a JADE implementation we summarize the architecture of an Experimental Air Traffic Simulator (EATS) [32] that we have developed under a JADE support. This simulator includes agents described in the last section as well as other two agents with particular purposes into a simulation environment: the *Configuration Agent* and the *Pseudopilot Agent*.

The *Configuration Agent* is required to define the set of initial simulation parameters (e.g. aircraft type, available routes, etc. The *Pseudopilot* Agent has been designed with a twofold purpose. First, it is a desk control that allows to an unique pilot-user (named pseudopilot) to have control over several aircraft. Second, it represents a graphical display, providing significant information about the state and intentions of surrounding traffic for each selected aircraft. This interface has been implemented as a separated agent (and not like an aircraft agent component), to centralize in a unique interface the access to each aircraft. Besides, it plays an important role (especially in the near future scenarios) to design and evaluate specific on-board man-machine interfaces like the CDTI cockpit display [50]. The CDTI allows seeing the surrounding traffic and, what is more relevant, the intentions of the surrounding aircraft. To access to a particular aircraft, a mouse click over the icon symbol is required. Once the aircraft is selected, it is placed at the central position of the *pseudopilot view window*, and the movement and the position of other aircraft are represented in relation to it. At the same time, the control window of the selected aircraft will be opened.

Then air-traffic controllers and pilots can interact with agents by means of two types of consoles. In one of them an Air Traffic Controller can monitor the positions of different aircraft and send several data instructions to a specific aircraft. In the other one a user pseudopilot that receives orders from the ATC (via voice or via data messages), carries out the necessary actions to fly the aircraft according to these orders. Besides, pseudopilot agent can be configured to automatically execute ATC mentioned data instructions.

Figure 13 shows screenshot of this application. It represents a view of the ATC interface constituted by and screen for displaying the traffic and a window console for interchanging data and instructions with a particular aircraft. Besides, in the same screenshot two aircraft control windows (A320 and Cessna) are deployed.

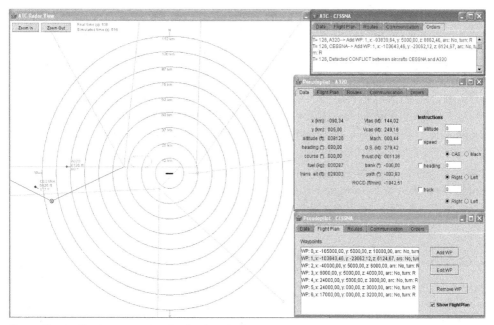

Figure 13. Application screenshot for the described scenario

To carry out communications between agents, a Communication class with specific methods has been designed. In particular, the air-ground communication between the aircraft agents and the ATC agent is carried out with the following messages:

a. Messages sent by the aircraft to the ATC: message to inform about the state vector and planed route, message to inform about the possible modification of the altitudes of the flight plan to initiate a continuous descent approach to the airport.

b. *Messages received in the aircraft from the ATC*: instruction messages (changing altitude, heading, speed, a flight plan waypoint, etc.) and messages of conflict detection with other aircraft. Starting from this nucleus of air-ground communications, new types of messages can be implemented in future EATS extensions with the purpose of establishing more complex negotiations between aircraft and ATC.

c. *Messages sent by the aircraft to others aircraft*: message to inform about the state vector. This air-air communication provides information to each aircraft about its surrounding air traffic.

Besides the previous communications, there are other communications involving the Meteorology Information Provider agent (to obtain atmospheric information) and the Airspace Recourse Provider agent (to request the available routes). Moreover, the communication between the aircraft agent and the pseudo pilot agent represents the communication between a physicals agent (the aircraft) and a man-machine interface like the CDTI.

Apart from above infrastructure for agent communications, the current prototype version implements the aircraft aerodynamic and simple agent making-decision mechanisms. Thus, aircraft agents are able to fly according a three dimensional flight plans or ATC vector instructions (i.e. heading, altitude and speeds orders). The aircraft aerodynamic model is based on a simplified point mass model that is described in [51]. In addition some navigation coordination tasks have been implemented on it (e.g. modifying arrival flight plan to perform a continuous descent and communicate this modification to the ATC agent via data message). In the same way, the ATC agent can detect missed separation conflict and provide primary conflicts resolution

Then, this architecture is intended as later extensions in order to add new algorithms (i.e. conflict detection and resolution algorithms, arrival sequence algorithms, etc.), air-air and air-ground negotiations protocols, human-machine interfaces and decision-making support systems, etc.

7. Conclusions

A summary description of a proposed conceptual model that represents TBO scenarios as a multi-agent system has been presented in this chapter. The aim of this design was to illustrate how current agent-oriented methodologies have been successfully applied to achieve highly structured representations of these scenarios and, therefore, they are a powerful design tool previous to a full implementation of future operational concepts.

A practical and formal methodological approach has been used to analyse and design the mentioned scenarios in a structured and consistent manner. By means of an iterative top-down modelling process the detailed agents architecture were designed based on capabilities, plans, events, and data structures.

The ATC view point was also described in this chapter through its inner architecture design. This architecture is oriented to execute several processes in order to plan, execute or modify trajectories in a coordinated way.

This approach has been based on guidelines provides by the recent Prometheus methodology. It has showed to be a suitable methodology for building models of next-generation ATM systems in the light of the following features:

- This approach achieves the goals of obtaining a highly structured model with several levels of abstractions. This structured nature allows a suitable integrity and consistence verification of the model on each of its three design phases: system specification, architecture design and detailed design.
- Also, Prometheus methodology provides proper guidelines for obtaining a system specification based on a goals hierarchical structure. These goals were identified from an organized set of scenarios that illustrate several aspects of the operational behaviour of the system. Goal at lowest level are used for defining required functionalities of the system as well as its main data, actions and perceptions.

- The overall system architecture combines information about roles of the air traffic entities with communication protocols that agents needs in order to improve their knowledge about the environment, agents states and intentions. Protocols are, also, a key aspect into the agent negotiation processes to achieve their respective goals.

- It illustrates how agent processes can be implemented by a set of several plans into agent. Plans are organized into several capabilities. Besides, this modular agent architecture based on plans allows a latter inclusion of new plans for implementing new procedures and functionalities. Moreover it is particularly important for obtaining robust software models.

- Finally, the model connects in a natural way those components that represent the dynamic perspective from those one than give a structural vision of the model. Protocols and processes, that model the dynamic behaviour, represent the core of the procedures. In the other hand capabilities are a high level representation of the systems required by ATC, aircraft and other air traffic entities to execute their task in a procedural manner.

After this first version of a simulation platform has been implemented and validated, new procedures, functionalities and underlying models will be included to be analysed as they are designed and included in the system following the described methodology. Furthermore, directions for future works include the extension of this conceptual model to gate-to-gate operations, as well as obtaining a full executable model for analytical simulation according to the described requirements.

Author details

José Miguel Canino
University of Las Palmas de Gran Canaria, Spain

Juan Besada Portas
University Polytechnic of Madrid, Spain

José Manuel Molina and Jesús García
University of Carlos III of Madrid, Spain

Acknowledgement

This work was funded by Spanish Ministry of Economy and Competitiveness under grant TEC2011-28626 C01-C02, and by the Government of Madrid under grant S2009/TIC-1485 (CONTEXTS).

8. References

[1]   Ljungberg M, Lucas A (1992) The OASIS air-traffic management system. Proceedings of the Second Paciffic Rim International Conference on Artificial Intelligence, PRICAI.

[2] Garcia, J. L. (1990). MAESTRO-A metering and spacing tool. American Control
     Conference, 1990 (pp. 502–507).
[3] Schubert, M. (1990). COMPAS system concept. The COMPAS System in the ATC
     Environment 19 p(SEE N 92-19041 10-04).
[4] NASA (2012), Overview of CTAS.
     http://www.aviationsystemsdivision.arc.nasa.gov/research/foundations/sw_overview.s
     html#overview
[5] van Gool M, Schoröter H (1999). PHARE Final Report, EUROCONTROL, Bruxelles.
     Availlable: http://www.eurocontrol.int/phare/gallery/content/public/documents/99-70-
     09pharefinal10.pdf
[6] RTCA (1995) Report of the RTCA Board of Directors' Select Committee on Free Flight,
     RTCA, Inc. Washington DC.
[7] Wilson, I. A. . (2007). 4-Dimensional Trajectories and Automation Connotations and
     Lessons learned from past research. Integrated Communications, Navigation and
     Surveillance Conference, 2007. ICNS'07 (pp. 1–10).
[8] Brooker P (2008) SESAR and NextGen: Investing in New Paradigms. Journal of
     Navigation, vol. 61, no. 2, pp. 195–208
[9] NASA (2006) Next Generation Air Transportation System (NGATS) Air Traffic
     Management (ATM)-Airspace Project. Reference Material. Available:
     http://cafefoundation.org/v2/pdf_tech/NASA.Aeronautics/PAV.NASA.ARMD.NGATS.
     pdf
[10] [10] NLR (2002) The NLR Air Traffic Control Research Simulator (NARSIM) Available:
     http://www.nlr.nl/documents/flyers/f075-07.pdf
[11] Weske R and Danek G (1993) Pseudo Aircraft Systems- A multi-aircraft simulation
     system for air traffic control research.   AIAA Flight Simulation Technologies
     Conference, Monterey, CA, 1993, pp. 234–242. Available:
     http://www.aviationsystemsdivision.arc.nasa.gov/research/foundations/pas.shtml
[12] FAA (2009) Target Generation Facility. http://hf.tc.faa.gov/capabilities/tgf.htm
[13] Vic D (2011) ATC interactive for the future of Air Traffic Control. URL:
     http://users.skynet.be/atcsim/
[14] Prevot T (2002) Exploring the many perspectives of distributed air traffic management:
     The Multi Aircraft Control System MACS. Proceedings of the HCI-Aero, 149-154.
[15] Clari M, Ruigrok R, Heesbeen B and Groeneweg J (2002) Research Flight Simulation of
     Future Autonomous Aircraft Operations. Proceedings of Winter Simulation Conference,
     San Diego, USA.
[16] Sweet D N, Manikonda V, Aronson, J S, Roth K, Blake, M (2002) Fast-time Simulation
     System for Analysis of Advanced Air Transportation Concepts· Proceedings of the
     AIAA Modeling and Simulation Technologies Conference, Monterey, CA.
[17] Callantine T J, Homola J, Prevot T, Palmer E A (2006) Concept Investigation via Air-
     Ground Simulation with Embedded Agents. Modeling and Simulation Technologies
     Conference and Exhibit, Reston, VA, USA.

[18] Callantine TJ, Palmer EA, Homola J, Mercer J Prevot T (2006). Agent-Based Assessment of Trajectory-Oriented Operations with Limited Delegation. 25th Digital Avionics Systems Conf., Portland, OR.

[19] Wangermann J P, Stengel R F (1998) Principled negotiation between intelligent agents: a model for air traffic management. Artificial Intelligence in Engineering, 12(3), 177-187.

[20] Wangermann J P, Stengel R F (1999) Optimization and Coordination of Multiagent Systems Using Principled Negotiation. Journal of Guidance, Control and Dynamics, 22(1), 43-50.

[21] Wangermann, J P and Stengel R F (1996) Distributed optimization and principled negotiation for advancedair traffic management. Proceedings of the 1996 IEEE International Symposium on. pp. 156-161.

[22] Harper K A, Mulgund S S, Guarino S L, Mehta A V and Zacharias G L (1999) Air traffic controller agent model for Free Flight. AIAA Guidance, Navigation, and Control Conference and Exhibit, pp. 288-301, Portland, OR.

[23] Stengel R F (1993) Toward intelligent flight control. Systems, Man and Cybernetics. IEEE Transactions on, 23(6), 1699-1717.

[24] Belkin B L, Stengel, R F (1987) Cooperative rule-based systems for aircraft control. 26th IEEE Conference on Decision & Control, Los Angeles, CA

[25] Stengel R F, Niehaus A (1989). Intelligent guidance for headway and lane control. Engineering Applications of Artificial Intelligence, 2(4), 307-314.

[26] Shandy S, Valasek J (2001) Intelligent agent for aircracft collision avoidance. AIAA Guidance, Navigation, and Control Conference, Montreal, Canada.

[27] Rong J, 2002, Intelligent Executive Guidance Agent For Free Flight. AIAA-2002-15 . Reno, NV

[28] Rong J, Ding Y, Valasek J, Painter J (2003) Intelligent system design with fixed-base simulation validation for general aviation. Proc. IEEE International Symposium on Intelligent Control, Houston, Texas.

[29] Painter J H (2002) Cockpit multi-agent for distributed air traffic management. En AIAA Guidance, Navigation, and Control Conference and Exhibit. Monterey, CA.

[30] Ding Y, Rong J, Valasek, J (2003) Automation Capabilities Analysis Methodology for Non-Controlled Airports. AIAA Modeling and Simulation Technologies Conference and Exhibit. Austin, Texas.

[31] Satapathy G, Manikonda V (2004). Agent Infrastructures for Modeling and Simulation of CNS in the NAS. En Fairfax, VA..

[32] Canino J M , García J, Besasa J., Gómez L (2008) EATS: An Agent-Based Air Traffic Simulator. IADIS International Journal of Computer Science and Information Systems, Vol. 3, Issue 2.

[33] Iglesias CA, Garijo M ,Gonzalez J C, Velasci J R (1996) A Methodological Proposal for Multiagent Systems Development Extending CommonKADS
        http://citeseernj.nec.com/

[34] Giunchiglia F, Mylopoulos J, Perini A(2002) The Tropos Software Development Methodology: Processes, Models and Diagrams. 2002 Autonomous Agents and Multi-Agent Systems (AAMAS 2002), Bologna, Italy

[35] Nwana H S, Ndumu D T, Lee L C, Collis J C (1999) ZEUS: A Toolkit and Approach for Building Distributed Multi-Agent Systems. J. M. Bradshaw, ed., Proceedings of the Third International Conference on Autonomous Agents (Agents '99), ACM Press, Seattle, USA, pp. 360-361.

[36] Wood M F, DeLoach S A, (2001) An Overview of the Multiagent Systems Engineering Methodology. Agent-Oriented Software Engineering, Volume 1957 of LNCS, Berlin: Springer, January 2001, 207-221.

[37] Wooldridge M, Jennings N R, Kinny D (2000) The Gaia Methodology for Agent-Oriented Analysis and Design. Journal of Autonomous Agents and Multi-Agent Systems, 3, (3), 285-312

[38] Pavón J, Gómez-Sanz J (2006) INGENIAS web site: http:// grasia.fdi.ucm.es/ingenias/, consulted at December 2006.

[39] Cernuzzi L, Rossi G (2002) On the evaluation of agent oriented modeling methods. En Proceedings of Agent Oriented Methodology Workshop, Seattle, November.

[40] Sturm A, Shehor O (2004) A Framework for Evaluating Agent-Oriented Methodologies. Lecture notes in computer science, 94-109.

[41] Luck M M, Ashri R, D'Inverno M (2004) Agent-Based Software Development, Artech House.

[42] Padgham L Winikoff M (2004) Prometheus: A methodology for developing intelligent agents. Lecture Notes in Computer Science, 174-185.

[43] Padgham L, Thangarajah J, Winikoff M (2008) Prometheus Design Tool, (System Demonstration), Proceedings of the Twenty-Third AAAI Conference on Artificial Intelligence (AAAI), Chicago, Illinois, USA.

[44] Luigi F, Caire G, Greenwood D (2007) Developing Multi-Agent Systems with JADE. Wiley Series in Agent Technology, Hardcover.

[45] Foundation for Intelligent Physical Agents –FIPA- (2007) Standard Status Specifications, http://www.fipa.org/repository/standardspecs.html

[46] Sorensen, John A, (2000), Detailed Description for CE-11 Terminal Arrival: Self Spacing for Merging and In-trail Separation, NASA Ames Research Center and NASA Langley Research Center, Moffett Field, CA and Hampton, VA.

[47] Huget M P (2004) Agent uml notation for multiagent system design. IEEE Internet Computing, 2004, vol. 8, pp. 63–71.

[48] Winikoff M (2004) JACK TM intelligent agents: An industrial strength platform", Bordini et al. , pp. 175–193.

[49] Foundation for Intelligent Physical Agents. 24 Communicative Act Library Specification, Version J. http://www.24.org/specs/2400037/

[50] Bone RS (2005) Cockpit Display of Traffic Information (CDTI) Assisted Visual Separation (CAVS): Pilot Acceptability of a Spacing Task During a Visual Approach. 6 USA/Europe Air Traffic Management R&D Seminar. Baltimore, MD.

[51] Glover W., Lygeros J. (2004). A Multi-Aircraft Model for Conflict Detection and Resolution Algorithm Evaluation. Technical Report WP1, Deliverable D1.3. Distributed Control and Stochastic Analysis of Hybrid Systems Supporting Safety Critical Real-Time Systems Design Project (HYBRIDGE). European Commission, Brussels.

Permissions

The contributors of this book come from diverse backgrounds, making this book a truly international effort. This book will bring forth new frontiers with its revolutionizing research information and detailed analysis of the nascent developments around the world.

We would like to thank Associate Professor Dr Tone Magister, for lending his expertise to make the book truly unique. He has played a crucial role in the development of this book. Without his invaluable contribution this book wouldn't have been possible. He has made vital efforts to compile up to date information on the varied aspects of this subject to make this book a valuable addition to the collection of many professionals and students.

This book was conceptualized with the vision of imparting up-to-date information and advanced data in this field. To ensure the same, a matchless editorial board was set up. Every individual on the board went through rigorous rounds of assessment to prove their worth. After which they invested a large part of their time researching and compiling the most relevant data for our readers. Conferences and sessions were held from time to time between the editorial board and the contributing authors to present the data in the most comprehensible form. The editorial team has worked tirelessly to provide valuable and valid information to help people across the globe.

Every chapter published in this book has been scrutinized by our experts. Their significance has been extensively debated. The topics covered herein carry significant findings which will fuel the growth of the discipline. They may even be implemented as practical applications or may be referred to as a beginning point for another development. Chapters in this book were first published by InTech; hereby published with permission under the Creative Commons Attribution License or equivalent.

The editorial board has been involved in producing this book since its inception. They have spent rigorous hours researching and exploring the diverse topics which have resulted in the successful publishing of this book. They have passed on their knowledge of decades through this book. To expedite this challenging task, the publisher supported the team at every step. A small team of assistant editors was also appointed to further simplify the editing procedure and attain best results for the readers.

Our editorial team has been hand-picked from every corner of the world. Their multi-ethnicity adds dynamic inputs to the discussions which result in innovative outcomes. These outcomes are then further discussed with the researchers and contributors who give their valuable feedback and opinion regarding the same. The feedback is then collaborated with the researches and they are edited in a comprehensive manner to aid the understanding of the subject.

Apart from the editorial board, the designing team has also invested a significant amount of their time in understanding the subject and creating the most relevant covers. They scrutinized every image to scout for the most suitable representation of the subject and create an appropriate cover for the book.

The publishing team has been involved in this book since its early stages. They were actively engaged in every process, be it collecting the data, connecting with the contributors or procuring relevant information. The team has been an ardent support to the editorial, designing and production team. Their endless efforts to recruit the best for this project, has resulted in the accomplishment of this book. They are a veteran in the field of academics and their pool of knowledge is as vast as their experience in printing. Their expertise and guidance has proved useful at every step. Their uncompromising quality standards have made this book an exceptional effort. Their encouragement from time to time has been an inspiration for everyone.

The publisher and the editorial board hope that this book will prove to be a valuable piece of knowledge for researchers, students, practitioners and scholars across the globe.

List of Contributors

Andrej Grebenšek
University of Ljubljana, Faculty of Maritime Studies and Transport, Portorož, Slovenia

Tony Diana
Division Manager, NextGen Performance, Federal Aviation Administration, Office of NextGen Performance and Outreach, ANG-F1, SW, Washington DC, USA

S.M.B. Abdul Rahman, C. Borst, M. Mulder and M.M. van Paassen
Control and Simulation Division, Faculty of Aerospace Engineering, Delft University of Technology, the Netherlands

Claudine Mélan
Shift-work and Cognition Laboratory, Toulouse University, France

Edith Galy
Research Centre in the Psychology of Cognition, Language, and Emotion, Aix-Marseille University, France

Kazuo Furuta, Kouhei Ohno and Taro Kanno
Department of Systems Innovation, the University of Tokyo, Japan

Satoru Inoue
Electronic Navigation Research Institute, Japan

R. Arnaldo, F.J. Sáez, E. Garcia and Y. Portillo
Universidad Politecnica de Madrid, Madrid, Spain

Luca Montanari and Roberto Baldoni
"Sapienza" University of Rome, Italy

Fabrizio Morciano and Marco Rizzuto
"Selex Sistemi Integrati" a Finmeccanica Company, Italy

Francesca Matarese
"SESM" a Finmeccanica Company, Italy

Tone Magister and Franc Željko Županič
SLOVENIA CONTROL, Slovenian Air Navigation Services, Ltd, Ljubljana, Slovenia

José Miguel Canino
University of Las Palmas de Gran Canaria, Spain

Juan Besada Portas
University Polytechnic of Madrid, Spain

José Manuel Molina and Jesús García
University of Carlos III of Madrid, Spain

Printed in the USA
CPSIA information can be obtained
at www.ICGtesting.com
JSHW011352221024
72173JS00003B/259